Ernst Schering Research Foundation Workshop 12
Interactions Between Adjuvants, Agrochemicals and Target Organisms

Ernst Schering Research Foundation Workshop

Editors: Günter Stock
Ursula-F. Habenicht

Vol. 6
Transgenic Animals as Model Systems for Human Diseases
Editors: E. F. Wagner, F. Theuring

Vol. 7
Basic Mechanisms Controlling Term and Preterm Birth
Editors: K. Chwalisz, R. E. Garfield

Vol. 8
Health Care 2010
Editors: C. Bezold, K. Knabner

Vol. 9
Sex Steroids and Bone
Editors: R. Ziegler, J. Pfeilschifter, M. Bräutigam

Vol. 10
Non-Genotoxic Carcinogenesis
Editors: A. Cochburn, L. Smith

Vol. 11
Cell Culture in Pharmaceutical Research
Editors: N. E. Fusenig, H. Graf

Supplement 1
Molecular and Cellular Endocrinology of the Testis
Editors: G. Verhoeven, U.-F. Habenicht

Vol. 12
Interactions Between Adjuvants, Agrochemicals and Target Organisms
Editors: P. J. Holloway, R. Rees, D. Stock

Ernst Schering Research Foundation
Workshop 12

Interactions Between Adjuvants, Agrochemicals and Target Organisms

P. J Halloway, R. T. Rees, D. Stock
Editors

With 52 Figures

Springer-Verlag Berlin Heidelberg GmbH

ISBN 978-3-662-02990-9 ISBN 978-3-662-02988-6 (eBook)
DOI 10.1007/978-3-662-02988-6

CIP data is applied

Viola-tions are liable for prosecution under the German Copyright Law.

Originally published by Springer-Verlag Berlin Heidelberg New York in 1994
Softcover reprint of the hardcover 1st edition 1994

Typesetting: Data conversion by Springer-Verlag

SPIN: 10471245 21/3130–5 4 3 2 1 0 – Printed on acid-free paper

Preface

Innovation is derived from a close interaction between basic research and industry. the most recent pharmaceutical successes based on molecular biological research provide examples of this. Innovation demands that researches learn about the developmental process, safety evaluation, production, and economics without downgrading their curiosity-driven research skills. This is an arduous task, but rewarding intellectually and perhaps also financially.

The main aim of our research proactiveness and this workshop is to encourage contact between such curiosity-driven researchers from academia and product-driven research and development. The result is, hopefully, an identification of frontier science and the means for its application in the field of the interaction between adjuvants, agrochemicals, and target organisms.

On reflection, there are many schools of thought on how to pursue the optimization of an agrochemical formulation, or to find the best adjuvant or adjuvant mixture to maximize the biological potential of an agrochemical against the specified target. There are, however, two extremes which must be reconciled. The first is the empirical approach. This is a costly method, formerly very much "spray" and "pray" or "spray" and "weigh", which in the past has predominated within the industry. The other is the purely rational approach, where the formulation or adjuvant is designed using only theoretical considerations, thereby integrating the findings of basic research.

Experience in the development of suitable formulations or adjuvants, a multidimensional problem, cannot easily be bought. It is hard task and a skill-requiring process which necessitates a critical ey coupled with a creditable knowledge of the biological processes involved

Fig. 1. The participants of the workshop

in addition to the physicochemical field. The "spray" and "weigh" image of the industry, we hope, has long been buried, and we are sure we will not let ourselves be fooled that once we know a compound's physical and chemical properties, we know exactly which adjuvant we require to maximize performance. Somewhere in between the extreme approaches there is a cost-effective, efficient compromise. The objective of the workshop was to establish the working practice across the disciplines and various fields of study in order to find that compromise.

Session of the workshop, dealt with physicochemical properties. Geoff Briggs and Richard Bromilow highlight the importance of the physicochemical properties of the active ingredient, allowing a more rational judgement about the potential for modification of biological activity by alteration of formulation variables. Here we would like to understand more fully the limiting factors imposed by the physicochemical nature of the a. i. and formulation components. They have been very willing to discuss the uptake and loss of the a. i. from the leaf surface with reference to choice of adjuvants. David Stock and Lorna Davies continue the physicochemical argument by presentin a case study using a model fungicide and explore the formulation adjuvant interaction which can significantly influence foliar penetration of the a. i.

Interaction with the target surface was the subject of session 2. Richard Falk surveys results of studies which have been conducted to investigate the influence of adjuvants and formulation variables on the foliar form of deposits and their distribution with respect to laf surface features. Peter Holloway examines the physicochemical interactions between adjuvants, spray droplets, and plant surfaces relevant to the modification of spray retention and coverage of the target, and presents examples of the influence of spray additives under identical track sprayer conditions (i. e. structure retention enhancement relationships). Bill Lankford reviews the penetration of compounds through the insect cuticle and proposes opportunities for the optimization of formulated insecticide products. This is the only paper that concentrates specifically on insects and mites and reopens the longstanding controversy over the pathways by which compounds penetrate to their site of action through the insect cuticle. Walter Steurbaut presents and reviews data about the influence of surfactants on the penetration and transport of fungicides into the pathogen, in addition to the influence of infection on adjuvant activity.

In session 3, methodology for evaluation was discussed. Peter Holloway sets the scene with a paper written from a theoretical viewpoint, suggesting methodology for monitoring specific adjuvant effects, and identifying possible modes of action. Per Kudsk and Solvejg Mathiassen present results on the assessment of formulations and adjuvants, the influence of adjuvant dose, and adjuvants as modifiers of biological activity under adverse application conditions. A section here is devoted to describe nonlinear regression techniques which may be used to interpret more easily the results generated from whole plant bioassay studies. Christian Gauvrit highlights some of the methodological problems encontered in the study of oil-based adjuvants. It leads to a discussion of the mechanisms involved, and how a better understanding of them would help in the design new oil adjuvants.

We hope that this workshop was for some the beginning of a longstanding relationship between those representing frontier scientific research and those in the agrochemical industry, while for others it will be a continuation of this process. Through the proceedings of this meeting, at the Schloß Cecilienhof in Potsdam, we have achieved what cannot be gained from scouring the literature: that is, the accumulation of the current knowledge as implied by the workshop's title, the exchange of ideas between experts in their field, the documentation of these ideas, and not only the planning for the future application of these ideas, but also the use of them as the basis to direct our own inhouse research, and finally to improve and strengthen our cooperation with all the invited participants.

R. T. Rees
D. Stock

Table of Contents

List of Contributors

G. G. Briggs
Schering Agrochemicals Limited, Chesterford Park, Saffron Walden,
Essex, CB10 1XL, UK

R. H. Bromilow
AFRC Institute of Arable Crops Research, Rothamsted Experimental Station,
Harpenden, Hertfordshire AL5 2JQ, UK

L. E. Davies
Schering Agrochemicals Limited, Chesterford Park, Saffron Walden,
Essex, CB10 1XL, UK

R. H. Falk
Division of Biological Sciences, Section of Plant Biology,
University of California, Davis, 95616 CA, USA

C. Gauvrit
Laboratoire de Phytopharmacie, INRA BV 1540,
21034 Dijon Cédex, France

P. J. Holloway
Department of Agricultural Sciences, University of Bristol,
AFRC Institute of Arable Crops Research, Long Ashton Research Station,
Bristol BS18 9AF, UK

P. Kudsk
Danish Institute of Plant and Soil Science,
Department of Weed Control and Pesticide Ecology, Flakkebjerg,
4200 Slagelse, Denmark

W. T. Lankford
NCH Europe, Codnor Gate Industrial Estate,Ripley,
Derbyshire, DE5 3NW, UK

S. K. Mathiassen
Danish Institute of Plant and Soil Science,
Department of Weed Control and Pesticide Ecology,
Flakkebjerg, 4200 Slagelse, Denmark

W. Steurbaut
Department of Crop Protection Chemistry,
Faculty of Agricultural and Applied Biological Sciences, University of Gent
Coupure 653, 9000 Gent, Belgium

D. Stock
Schering Agrochemicals Limited, Chesterford Park, Saffron Walden, Essex,
CB10 1XL, UK

1 Influence of Physicochemical Properties on Uptake and Loss of Pesticides and Adjuvants from the Leaf Surface

G. G. Briggs and R. H. Bromilow

1.1 Introduction

The literature on the effects of adjuvants on the uptake of chemicals by plants, insects, isolated cells, micro-organisms and skin is very complex. The physical properties of both the active ingredient and the adjuvant clearly play a part, although its exact nature is generally far from clear.

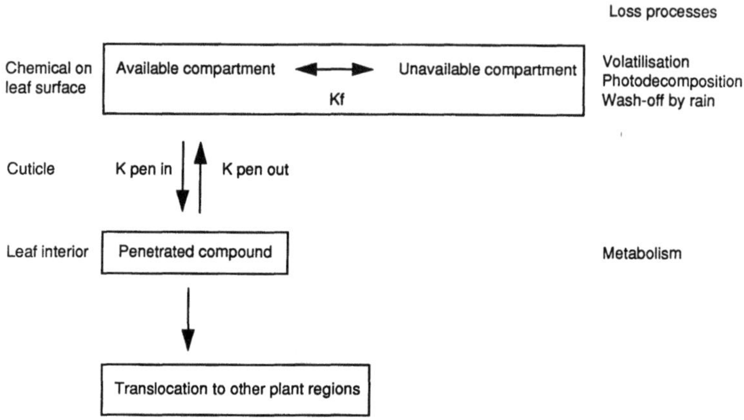

Fig. 1. General model of foliar uptake of chemicals by plants. (Adapted from McCall 1988)

Which properties are relevant? The description of percutaneous absorption by Higuchi (1977) offers some guidance. For a solid drug the four steps involved in penetration of a compound are: (a) release of a solid to solution, (b) uptake from solution into the skin, (c) clearance into the sub-barrier environment, and (d) longer distance movement and metabolism. General empirical rules governing penetration are: (a) non-polar solutes are absorbed better than polar ones, (b) lipoidal agents are more easily delivered than non-lipids, (c) solids with low melting point are more easily absorbed than those with high melting point, and (d) liquids are better absorbed than solids. These generalisations could equally be made for foliar penetration and are formalised in the model of foliar uptake (Fig. 1) adapted from McCall (1988). In this model chemical on the leaf surface is considered to be in states in which it is either available or unavailable for uptake. Volatilisation is a possible loss process from these compartments. In the field, photodecomposition and wash-off by rain are important possible loss processes in the absence of uptake.

Penetration into the leaf occurs from the available compartment, but reverse movement is possible and may become appreciable when the surface concentration is lower than that inside the plant. This factor was included to account for the changes of concentration in plants of tridiphane, which is both volatile and a rapid penetrator of plants. The existence of such outward movement is supported by good experimental evidence (Schering Agrochemicals Limited, unpublished reports). Disease control was observed on untreated plants adjacent to plants growing in soil treated with a volatile systemic fungicide. The conditions were such that the chemical could have moved only via the transpiration stream of the treated plants, and thereafter was released by volatilisation from the leaves. The chemical once taken up by the roots or leaves may move to other parts of the plant and also then be subject to metabolism.

The uptake of chemicals by plant roots and movement in the xylem or phloem, either following root uptake or stem injection, is now well understood on the basis of physical properties. This work is discussed in Sect. 1.2. The search for a similar explanation of foliar uptake based on physical properties follows in Sects. 1.3–1.6.

1.2 Uptake by Roots and Systemic Movement in Plants

Uptake of non-ionised pesticides was studied using barley plants grown in nutrient solution and then transferred to solution containing ^{14}C-labelled test compound at 5 μM concentration (Briggs et al. 1982). Distribution of the parent compound was measured following extraction and thin layer chromatography. Distribution of the compound between the roots and the external solution was defined by the root concentration factor (RCF):

$$\mathrm{RCF} = \frac{\text{concentration in fresh roots}}{\text{concentration in nutrient solution}}$$

RCF was related to log P_{oct} (octanol-water partition coefficient) by the simple relationship:

$$\log(\mathrm{RCF} - 0.82) = 0.77 \log P_{oct} - 1.52 \tag{1}$$

The factor 0.82 represents compound dissolved within the aqueous phase of the root cells and is assumed to be constant for all compounds. The slope term, 0.77 log P_{oct}, represents partitioning into the plant material and is the dominant term for log $P_{oct} > 1.5$. Such partitioning is reversible and was measured independently using root material that had been frozen, ground and freeze dried. Similar findings have been reported for plant leaves and stems (Briggs et al. 1983; Uchida 1980; Barak et al. 1981); the transport of compounds in the xylem away from the point of entry is akin to a crude type of reverse-phase chromatography. The above relationship applies reasonably well for a range of non-woody plant species, indicating a general similarity in their sorbing capacity.

Transport of non-ionised compounds from the roots to the above-ground parts requires passage across the endodermal cells that surround the vascular tissues, which are sealed tightly together by the lignified and suberised Casparian Strip. Compounds of log P_{oct} ca. 1.8 cross the endodermis with almost the same efficiency as water, but more polar and more lipophilic compounds cross this barrier much less readily. This process can be described mathematically by the transpiration stream concentration factor (TSCF):

$$\mathrm{TSCF} = \frac{\text{Concentration in xylem sap}}{\text{Concentration in the nutrient solution}}$$

$$\mathrm{TSCF} = \frac{\text{Amount accumulated in foliage (corrected for degradation)}}{\text{Volume of water transpired} \times \text{concentration in nutrient solution}}$$

The data were modelled by the gaussian curve:

$$\mathrm{TSCF} = 0.784 \exp - [\log P_{oct} - 1.78)^2/2.44] \qquad (2)$$

Uptake of ionised compounds is more complex. Distribution of weak acids is determined by their pK_a, and by the pH of the plant compartments and that of the nutrient or soil solution. Undissociated acids move much more readily through membranes than their anions because of the charge and the high polarity of the latter. Accordingly, weak acids are taken up most efficiently from solutions of lower pH and then accumulate within compartments of high pH, this being the so called "ion trap" effect. Such processes also influence the extent of transport within the plant, which is via both xylem and phloem for acidic compounds; partitioning into the plant material is negligible for all except lipophilic (log $P_{oct} > 4$) acids (Briggs et al. 1987).

Subsequent transport of a compound to the shoot depends on the amount accumulated in the roots. However, the efficacy of transport from the root cortex to the xylem (via the endodermis and the Casparian

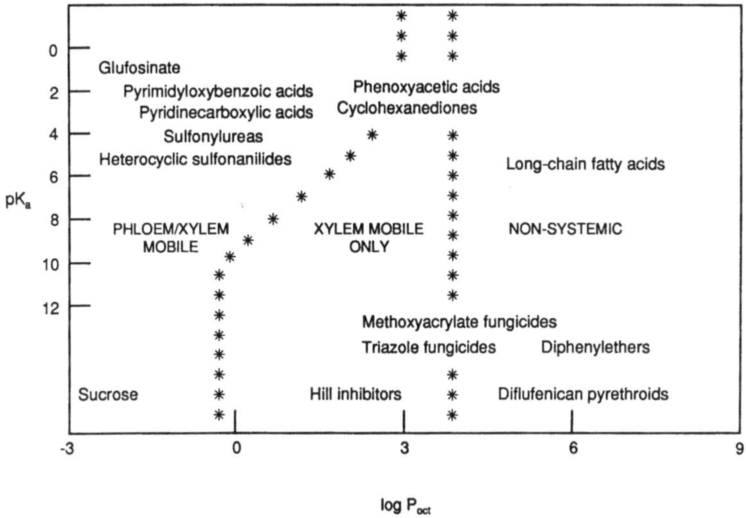

Fig. 2. Influence of pK_a and log P_{oct} on the phloem and xylem mobility of chemicals in plants. (Adapted from Bromilow and Chamberlain 1991)

Strip) is rather low, at less than 10%, which indicates the difficulty with which weak acids, substantially ionised at physiological pH, are able to cross membranes. Compounds entering leaves appear to move in similar ways, whatever the route of entry. For example, Bromilow and Chamberlain (unpublished results) observed similar transport efficacies to phloem sinks for a range of substituted phenoxyacetic acids in the castor bean plant, after application by petiole injection of aqueous solutions, to those in barley following zone application to a leaf using aqueous acetone solutions containing NPE8 surfactant [a nonylphenol with an average ethylene oxide (E) content of 8]. Movement in plants can be predicted from pK_a and log P_{oct} (Fig. 2; adapted from Bromilow et al. 1990; Bromilow and Chamberlain 1991). Lipophilic compounds (log $P_{oct} > 4$) are not systemic, whilst compounds of intermediate lipophilicity (log P_{oct} – 0.5 to + 3.5) move in the xylem; weak acids additionally move in the phloem to sinks. A key point to appreciate is that, contrary to expectations based on sucrose which requires active transport to enter the phloem, most compounds enter the phloem freely but, in the absence of a mechanism for retention such as the "ion trap", leave it equally readily. Movement is thus seen to be in the direction of the much larger transpiration stream flow.

1.3 Physical Properties Affecting Uptake into Leaves

The previous section shows that uptake of chemicals by roots and their subsequent movement in plants is well understood from a knowledge of two readily estimated physical properties. A large body of experimental data is rationalised, and chemical structures can be predictably modified to obtain a required behaviour. In contrast, there is no comparable rationale for uptake by leaves. Most compounds penetrate most plants poorly when applied alone, although there are exceptions. Baker et al. (1992) were unable to find correlations between uptake and melting point, log P_{oct} or water solubility of a diverse group of pesticides, in contrast to the dependence on log P_{oct} reported by Chamberlain et al. (1987) for two sets of closely related chemicals.

In terms of the McCall model (Fig. 1), most compounds are unavailable for uptake and require an adjuvant for uptake to occur. The most obvious explanation is that the adjuvant increases the fraction available for uptake. It seems unlikely that the intrinsic permeability of the compound, or the leaf surface, are changed by the adjuvant at the rates commonly used, although this is a subject for debate.

Can we understand more about the behaviour on leaf surfaces of active ingredients by understanding the behaviour of the adjuvants themselves? A major difficulty is the lack of data on the physical properties of adjuvants, which are seldom described in the same way as active ingredients. The classification of adjuvants as "inerts" because they have no biological activity, at least at the low rates used in traditional formulations to affect the mechanical aspects of spraying, seems to have formed the conclusion that they also have no chemistry.

A number of properties are needed to use the McCall model as a framework to assess the behaviour of adjuvants on and in plants. An estimate of volatility is clearly required. The surface form of the adjuvant affects its own uptake as well as that of the active compound. The log P_{oct} of an active ingredient influences the choice of adjuvant (Stock et al. 1993) and, as described in Sect. 1.2, the degree and direction of movement after penetration. Work on blood-brain barrier penetration

(Young et al. 1988) and skin penetration (Tayar et al. 1991) has used the parameter >λογ ΠΔ log *P*, where:

$$\Delta \log P = \log P_{oct} - \log P_{alk} \quad (3)$$

and P_{alk} is the alkane/water partition coefficient, generally hexane or cyclohexane.

The important feature of Δ log *P* is that it is the difference between the > log PΔ log *P* values which is important, rather than their absolute values. For example, extending an alkyl chain would leave the log *P* value essentially unchanged. Since the epicuticular wax is essentially hydrocarbon in character (Baker 1982), this parameter is clearly relevant to foliar uptake. Water solubility is a commonly measured property of active ingredients, and it is also required for adjuvants.

Estimates of these properties, given in Tables 2–5, are very approximate. Some adjuvant materials are complex mixtures and values for the mixture may be modified by the presence of components having a range of properties. However, the values are useful as a general guide and are reasonably self-consistent. For example, the water solubilities of the ethoxylated surfactants, calculated from calculated log P_{oct} values, are close to their critical micelle concentrations (see Table 5).

The convention used to name alcohol ethoxylates is to give the alcohol chain followed by the number of $-CH_2CH_2O-$ units. Ethoxyethanol, $CH_3CH_2OCH_2CH_2OH$, is therefore C_2E_1.

1.3.1 Volatility

Hartley and Graham-Bryce (1980) describe a simple method to validate experimental estimates of the vapour pressure of compounds by converting their structure to a hydrocarbon, using carbon equivalents for hetero-atoms. If the vapour pressure reported for the compound is greater than that of the equivalent hydrocarbon, the result is dubious. A more extensive list of factors has been derived which enables order of magnitude estimates of vapour pressure for most compounds (Briggs, unpublished results).

To make interpretation simple, volatility is expressed as the equivalent hydrocarbon (EH), which is the number of carbon atoms in the simple benzene or alkane with the same expected boiling point and vapour pressure. As a simple example, chlorine has a carbon equivalent of 2 and hence an EH of chlorobenzene is 8. The boiling point of chlorobenzene is 132°C, similar to that of the C8 hydrocarbons *p*-xylene (135°C), octane (125°C), dimethylcyclohexane isomers (120°–125°C) and allylcyclopentane (125°C). Highly branched structures, such as 2,2,4-trimethylpentane (boiling point 99°C) have lower boiling points, which can be taken into account by using branching factors or by using suitable partial structures as the starting point for a more complex molecule.

For example, oxygen in an ether link and silicon are both equivalent to 1 carbon. Hence, for the ethoxylated trisiloxanes contained in Silwet L-77 (mean ethylene oxide content 8, oligomer range 2–14):

$$\begin{array}{c} CH_3 \\ | \\ (CH_3)_3SiOSiOSi(CH_3)_3 \\ | \\ CH_2CH_2CH_2(OCH_2CH_2)n\text{-}OCH_3 \end{array}$$

when n = 2, the equivalent hydrocarbon is C15 + 3(Si) + 5(ether O), i.e. EH = 23. Reference to the boiling point of octamethyltrisiloxane, indicates that the estimate for the siloxane chain is about 3 carbons too high, so that the corrected EH = 20. This 2E oligomer elutes on gas liquid chromatography close to an n-C_{19} hydrocarbon standard (Zabkiewicz et al. 1993). EH values in subsequent tables were similarly derived.

The expected behaviour of compounds with EH 10–30 is summarised in Table 1. Any compound with EH 25 undergoes appreciable vapour loss over 1 day. At EH = 30 the possible vapour loss is only a few grams per hectare per day, leading to observable vapour effects in the growth room and glasshouse for highly biologically active compounds, but unlikely to affect performance of adjuvants at their standard rates of use.

1.3.2 Partition Coefficients

The Pomona College "Clog P programme" and the associated database were used as the source of the octanol-water partition coefficients in the

Table 1. Votalisation rate and vapour pressure of equivalent hydrocarbons (EH)

Hydrocarbon	Vapour pressure (25 °C)		Volatilisation rate (kg/ha per day)	$t_{1/2}$ on leaf surface for 1 kg/ha
	Pascal	mm Hg		
EH 10	400	3	68 000	<<< 1 min
EH 15	9	7×10^{-2}	1 400	< 1 min
EH 20	0.1	1×10^{-3}	30	30 min
EH 25	4×10^{-3}	3×10^{-5}	0.5	1 day
EH 30	8×10^{-5}	6×10^{-7}	0.01	50 days

following sections. The calculated values were modified in some cases to give consistent values. Those quoted in Table 5 for the long-chain ethoxylated surfactants are no more than approximations.

Alkane-water partition coefficients are available in the database for some compounds. Others were estimated using the fragment values of Leahy et al. (1992).

1.3.3 Solubilities

The water solubility of organic liquids (in moles/l) was estimated from:

$$\log \text{water solubility} = -\log P_{oct} \tag{4}$$

and for solids, of melting point T_m C, from:

$$\log \text{water solubility} = 0.6 - 0.01\ T_m - \log P_{oct} \tag{5}$$

A very approximate estimate of solubility (as moles per litre) in moderately polar organic solvents, such as octanol, esters or aromatic hydrocarbons, is given by:

$$\log \text{organic solubility} = 0.6 - 0.01\ T_m \tag{6}$$

Where strong H-bond donor or acceptor functionalities are present in the solute, solubility in ketones or chloroform, respectively, may be much higher than in other solvents. The functionalities concerned can be identified from the data given in Leahy et al. (1992). Alkanes are exceptional in their very low water content at saturation and are poor solvents for compounds with polar functionalities. Solubility in alkanes is given approximately by:

$$\text{log alkane solubility} = 0.6 - 0.01\ T_m - (\log P_{oct} - \log P_{alk}) \qquad (7)$$

The relationships described above are not precise but give reasonable estimates for many pesticides using data from Worthing and Hance (1991).

The key property is melting point, high values arising from high lattice energies which decrease the solubility in all solvents. Partition coefficients are effectively ratios of solubilities so that it is possible to have lipophiliccompounds with high melting points which are poorly soluble in organic solvents and relatively polar liquids with good organic solvent solubility. Table 2 summarises these relationships between solubilities, partition coefficients and melting points for compounds of molecular weight 250. The importance of the melting point is clear. The data in Table 2 can also be used to indicate formulation possibilities for active ingredients. High concentrations are the only practical ones for compounds with use rates of 1 kg/ha. The data in Table 2 show that such compounds with melting points above 100°C cannot be made into either solvent- or water-based formulations unless they are very polar neutral molecules or salts, when an aqueous based solution concentrate is possible, or unless powerful polar organic solvents are used for EC formulations.

As can be seen from Table 2, the alkane solubility for compounds with large log *P* values is too low to permit entry into plants via the epicuticular wax, and an aqueous route for uptake (ill-defined and lacking experimental verification) must be involved. For glyphosate salts, with water solubility greater than 200 g/l and estimated alkane solubility of 0.000000000001 g/l, there is no other way.

The properties of solvents and adjuvants commonly used in formulations or experimental studies of foliar uptake are given in Tables 3–5. The consequences for their behaviour on and in plants are discussed in the next section.

Table 2. Physical properties of compounds with molecular weight 250 and solubility and organic solvents, alkanes and water

Melting point °C	Moderately polar organic solvents	Solubility g/l										
		Alkanes				Water						
		$\Delta \log P$				$\log P_{oct}$						
		1	2	3	4	–2	–1	0	1	2	3	4
oil	500	100	10	1	0.1	>500	>500	>500	100	10	1	0.1
50°	300	30	3	0.3	0.03	>500	>500	300	30	3	0.3	0.03
100°	100	10	1	0.1	0.01	>500	>500	100	10	1	0.1	0.01
150°	30	3	0.3	0.03	0.003	>500	300	30	3	0.3	0.03	0.003
200°	10	1	0.1	0.01	0.001	>500	100	10	1	0.1	0.01	0.001
250°	3	0.3	0.03	0.003	0.0003	300	30	3	0.3	0.03	0.003	0.0003
300°	1	0.1	0.01	0.001	0.0001	100	10	1	0.1	0.01	0.001	0.0001

Examples: (a) Solubility in organic solvents falls from 500 to 10 g/l as melting point increases from an oil at ambient temperature to 200°C; (b) When $\Delta \log P = 4$, even oils dissolve only in alkanes at 0.1 g/l; (c) water solubility of fluquinconazole, $\log P_{oct} = 3.2$ and melting point = 191 °C, is 0.001 g/l.

1.4 Behaviour of Adjuvants on Leaf Surfaces

1.4.1 Volatilisation and Penetration

At applications of 1 l/ha (10 $\mu g/cm^2$) the common solvents listed in Table 3 are all highly volatile, with nearly complete loss expected in minutes. Their influence on the penetration of active ingredients can therefore be of only short duration. The data of Krenek et al. (1987) on the phytotoxicity of solvents when applied at rates of 10–80 l/ha indicate that phytotoxicity is greatest with the less lipophilic and less volatile aromatic hydrocarbon, alcohol, ketone and ester solvents which presumably penetrate well despite their relatively high volatility.

The fatty acids, alcohols and esters listed in Table 4 span a wide range of properties. Simple esters such as methyl oleate should be quite volatile, as observed by Urvoy et al. (1992). Despite its lipophilicity methyl oleate was shown to penetrate maize steadily, in contrast to triolein which penetrated for a short time and then ceased. The difference probably lies in the potential for rapid hydrolysis of methyl oleate after penetration, resulting in a continuous concentration gradient from the surface deposit. The high lipophilicity of the esters or of oleic acid itself (log P_{oct} 7) preclude any transport other than that of metabolites away from the treated area (Fig. 2). Nonanoic acid, being marginally the most phytotoxic of the C_8–C_{10} fatty acids, is used to give short term scorch symptoms in otherwise slowly acting glyphosate formulations (Arnold et al. 1993). Rapid penetration of these acids must therefore be possible. Lauric acid (C_{12}) was much less phytotoxic, probably because it is above the log P_{oct} limit for good movement in plants (Fig. 2).

Table 5 lists the estimated properties of the commonest surfactant adjuvants, the ethoxylated alcohols and phenols. Other than the short-chain alcohols with an E content of 1–5 (little used in practice), volatilisation is unlikely to be significant. The C_8–C_{10} alcohol ethoxylates potentially are mobile in plants after penetration, as indicated by log P_{oct}, but the C_{12} and higher alcohol ethoxylates and the alkylphenol ethoxylates with less than about 15E would not be expected to be mobile, as shown by Stock and Holloway (1993). Although the E15–20 and higher ethoxylates potentially are mobile, relevant data for long-chain ethoxylates do not appear in the literature. Since they penetrate slowly with half-lives for uptake of ca. 2 days, these being longer than

Table 3. Physical properties of solvents commonly used in experimental and commercial formulation

Solvent	EH	Boiling point (°C)	log P_{oct}	log P_{alk}	Water solubility (g/l)
Xylene	8	140	3.1	~3.1	0.08
Chlorobenzene	8	130	2.8	~2.8	0.2
C_{12} paraffin	12	216	6.6	6.6	0.00005
C_{15} paraffin	15	271	8.1	8.1	0.000001
Solvesso 200[a]	14–17	226–279	4–6	4–6	~0.005
Heptyl acetate	11	~190	3.2	3.0	0.1
Di-isononylphthalate	>30	>400	9.5	9.3	<10–7
Acetone	5	56	–0.2	–0.7	Miscible
Acetophenone	11	202	1.6	1.1	3.0
Isophorone	12	213	2.2	1.7	0.9
Dimethylformamide	9	153	–0.1	–2.7	Miscible
N-Methylpyrrolidone	13	~200	–0.7	–2	Miscible
Isopropanol	7	82	0	–1.8	Miscible
Octan-1-ol	12	197	3.1	1.3	0.1
Ethylene glycol	12	198	–1.9	–4.7	Miscible
Glycerol	18	291	–1.8	–5.7	Miscible
Water	7	100	–1.4	~–7	Miscible
Dimethyl sulfoxide	11	189	–1.3	–4.4	Miscible
Isopropylamine[b]	5	35	0.3	–0.9	Miscible

[a] Alkylnaphthalenes.
[b] pK_a 10.6.

the metabolic half-life in plants of ca. 1 day for C12E8 and NPE5.5 indicated by the data of Silcox and Holloway (1986), experimental verification of their mobility would be difficult using foliar treatments.

A feature of the ethoxylate chain is the small decrease in log P_{oct} (and hence a small increase in water solubility) with increasing ethoxylation of ca. 0.1 per E unit. Similar small changes, ca. 0.15 per E, have been observed for dewaxed plant cuticle/water partition coefficients (Schönherr et al. 1990). As would be expected from the ester nature of the cuticle matrix, octanol is probably a reasonable model for the cuticle.

Table 4. Physical properties of fatty acid esters, fatty acids and alcohols

Adjuvant	EH	Boiling point (°C)	log P_{oct}	log P_{alk}	Water solubility (g/l)	Melting point (°C)
Methyl laurate	15	262	5.4	5.2	0.001	Oil
Methyl palmitate	19	~350	7.4	7.2	0.00001	30
Methyl stearate	21	377	8.4	8.2	0.000001	39
Methyl oleate	21	~370	8.0	7.8	0.000001	20
Butyl oleate	24	~420	9.5	9.3	0.0000001	Oil
Octyl oleate	28	~500	>10	>10	$<1 \times 10^{-9}$	Oil
Tracetin[a]	15	258	–0.2	–0.4	100	Oil
Triolein	>>30	>>500	>10	>10	$<1 \times 10^{-12}$	Oil
Octanoic acid[b]	14	240	3.0	0.6	0.2	16
Nonanoic acid	15	255	3.5	1.2	0.05	32
Decanoic acid	16	270	4.0	1.8	0.01	32
Lauric acid C12	18	~310	~5.0	3.0	0.001	46
Palmitic acid	24	~440	~7.0	5.6	0.00001	64
Decanol	14	231	4.1	2.4	0.01	Oil
Dodecanol	16	259	5.1	3.5	0.001	26

[a] Glyceryl triacetate.
[b] Fatty acid pk_a 4.8.

In contrast to the small change in log P_{oct}, log P_{alk} decreases sharply with ethoxylation by ca. 0.4 per E unit. This is in accord with only the lower ethoxylates, up to about 6E reported in manufacturers' literature, being soluble in hydrocarbons. Increasing ethoxylation thus creates molecules with unchanged lipophilicity on the log P_{oct} scale, but a much decreased lipophilicity on the log P_{alk} scale. log P therefore increases. The consequence should be a much lower penetration rate into plants for the long-chain higher ethoxylates, which are too polar to penetrate readily via the epicuticular wax, but at the same time too water insoluble ($1 \times 10^{-3} M$ or less) to penetrate by an aqueous route. The literature on uptake of ethoxylates (Silcox and Holloway 1986;Stock and Holloway 1993) indicates that the lower ethoxylate penetrate well, but that higher ethoxylates penetrate much more slowly and in a different manner.

Table 5. Estimated physical properties of ethoxylated alcohols and allyl phenols

Adjuvant	EH	log P_{oct}	Water solubility mole/l[a]	log P_{alk}
C_8 alcohol				
E2	19	2.7	2×10^{-3}	1.1
E5	28	2.4	4×10^{-3}	–0.1
E10	>30	1.9	1×10^{-2}	–2.1
E15	>30	1.4	4×10^{-2}	–4.1
E20	>30	0.9	1×10^{-1}	–6.1
C_{10} alcohol				
E5	30	3.4	4×10^{-4}	1.1
E10	>30	2.9	1×10^{-3}	–0.9
E15	>30	2.4	4×10^{-3}	–2.9
E20	>30	1.9	1×10^{-2}	–4.9
C_{12} alcohol				
E5	>30	4.4	4×10^{-5}	2.3
E10	>30	3.9	1×10^{-4}	0.3
E15	>30	3.4	4×10^{-4}	–1.7
E20	>30	2.9	1×10^{-3}	–3.7
C_{18} alcohol				
E5	>30	7.4	4×10^{-8}	5.9
E10	>30	6.9	1×10^{-7}	3.9
E15	>30	6.4	4×10^{-7}	1.9
E20	>30	5.9	1×10^{-6}	–0.1
Octylphenol				
E5	>30	4.1	7×10^{-5}	1.6[b]
E10	>30	3.6	2×10^{-4}	–0.4[c]
E15	>30	3.1	8×10^{-4}	–2.4
E20	>30	2.6	2×10^{-3}	–4.4
Nonylphenol				
E5	>30	4.4	4×10^{-5}	2.5
E10	–	3.9	4×10^{-4}	0.5
E15	–	3.4	4×10^{-4}	–1.5
E20	–	2.9	1×10^{-3}	–3.5

[a] Water solubility approximately equals critical micelle concentration.
[b,c] Literature measured values 1.6 and –0.6.

Table 5. Continued

Adjuvant	EH	log P_{oct}	Water solubility mole/l[a]	log P_{alk}
Silewt L-77 components[d]				
E2	21	4.6	2×10^{-5}	3.5
E5	30	4.3	5×10^{-5}	2.0
E8	>30	4.0	1×10^{-4}	0.5
Sylgard 309[e]				
E8	>30	4.0	1×10^{-4}	0.8
Amine ethoxylates[f]				
AM16E20	>>30	4.8	2×10^{-5}	–5
Sorbitan monolaurate				
Tween E20	>>30	1.2	5×10^{-2}	<–8
Sodium dodecyl sulphate				
(Melting point 205 °C)	>>30	0.6	4×10^{-3}	<–6

[d] Ethoxylated heptamethyl-3-hydroxypropyltrisiloxane with terminal methyl ether.
[e] Same siloxane with terminal acetyl ester.
[f] pK_a of amine ca. 8; data for free base.

1.4.2 Adjuvants as Solubilising Agents

The solubility data in Table 2 indicate the solubilising capacity of the many types of adjuvant materials for active ingredients. Regarding the adjuvants as organic solvents, equivalent in solvent power to an aromatic hydrocarbon or aliphatic alcohol, indicates that active ingredients that are liquids or are solids with low melting point could be partially solubilised by a 1:1 ratio with adjuvant, but that higher melting solids would need more adjuvant.

An additional feature of the ethoxylated surfactants is their increased water retention with increasing ethoxylation. This is humidity dependent. The data of Anderson and Girling (1983) and Stevens and Bukovac (1987) indicate that at 80% relative humidity water retention is approximately 1 mol H_2O for a 5E chain, 5 for 10E, 12 for 15E and 20 for 20E; these values increasing two- to three-fold at 90% humidity. The short chain ethoxylates (to ca. E7) are therefore relatively hydrophobic compounds, while the long-chain compounds contain at least 30 *M* H_2O (except at low humidities) and could therefore solubilise polar compounds in H_2O quite effectively. At low humidities the solubilising capacity for hydrophobic active ingredients would not change, but for water-soluble ones it would decrease dramatically, particularly as the long-chain ethoxylates tend to be solids at low humidity.

1.4.3 Leaf Surface Reactions of Adjuvants

Sulphonate salts are solids with high melting points. The data of Anderson and Girling (1983) indicate that crystallisation of sodium dodecyl sulphate occurs at 90% relative humidity. Calcium salts are generally much less soluble than sodium salts, in part because they have much higher melting points. Since calcium and, less frequently, sodium sulphonates are components of EC formulations, it is possible that the uptake measured from such EC formulations has been influenced by crystallisation of these salts. It is possible that uptake of the cation could occur which, in the presence of even a weakly basic active ingredient, could lead to the formation of insoluble salts. The antagonism of glyphosate by calcium salts due to the formation of insoluble calcium salts and its prevention by sulphate is a good example of competitive salt formation (Nalewaja and Matysiak 1993). One would expect that the isopropylamine salt of glyphosate would lose isopropylamine by volatilisation in the presence of the amine ethoxylates used in commercial formulations. Although a typical C16E20 amine is 400 times less basic than isopropylamine, isopropylamine is many million times more volatile. Many experiments are reported in the literature in which precisely defined adjuvants are used under controlled environmental conditions with pure radiolabelled compounds which are then mixed with commercial formulations of unspecified composition. These formulations have a chemistry which cannot be ignored!

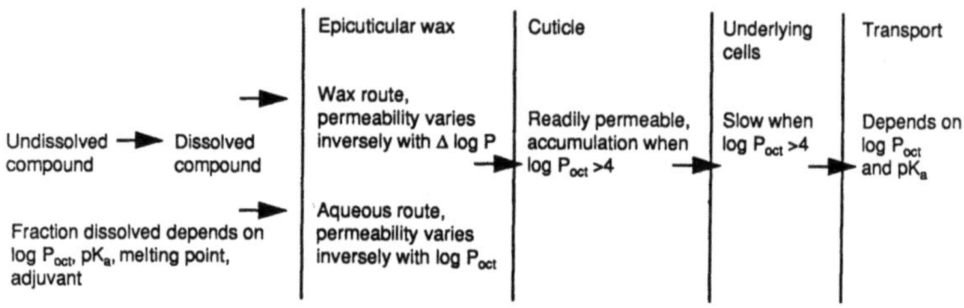

Fig. 3. Factors influencing foliar uptake processes

1.5 General Outline of Foliar Uptake

Based on the uptake properties of many adjuvants and on the general literature on uptake of active ingredients, the following view of uptake by leaves is proposed in Fig. 3.

To penetrate readily, a compound must be either a liquid or in solution. Solubility is influenced by the physical properties of the compound, the solvent properties of any adjuvant present and the adjuvant concentration. Once the compound is dissolved, the melting point ceases to be important. Penetration into the wax depends on $\Delta \log P$ and is poor for polar compounds whatever the value of $\log P_{oct}$. Since $\Delta \log P$ is positive for most compounds, those entering the wax then move readily into the cuticle because they are more strongly absorbed by the octanol-like cuticle. The instinctive expectation is that lipophilic compounds should accumulate in the epicuticular wax, but experimentally this is not observed; entry into the wax is the limiting factor, not further penetration. Uptake via the aqueous route occurs for compounds with high water solubility.

Retention in the cuticle is appreciable for compounds with high $\log P_{oct}$, with further penetration to the underlying cells and transport also being restricted. Complex patterns of behaviour are observed if the mixture of compound and adjuvant contains much solid material, or if the two components have different rates of penetration. These principles are illustrated in Figs. 4–9, which show the consequences of varying $\Delta \log P$, $\log P_{oct}$ and melting point. The compounds chosen exemplify work from the literature or this paper. It is not possible to find examples of all possible combinations of properties because (a) most experimental

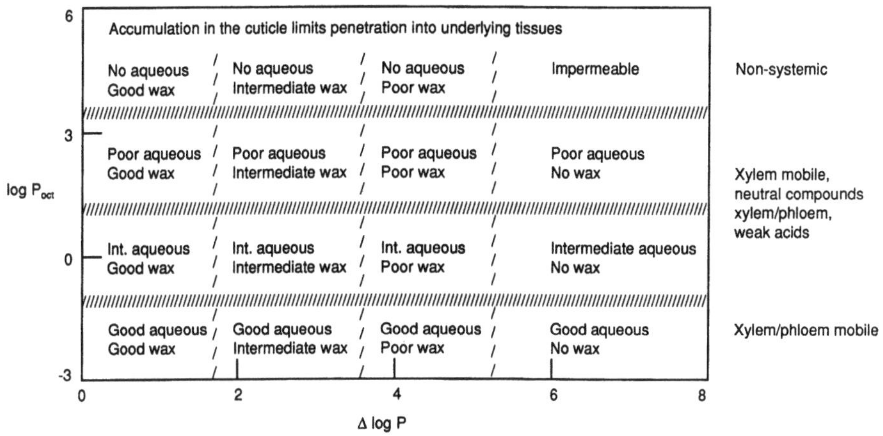

Fig. 4. Effect of log P_{oct} and Δ log P on foliar uptake via epicuticular wax or aqueous pathways

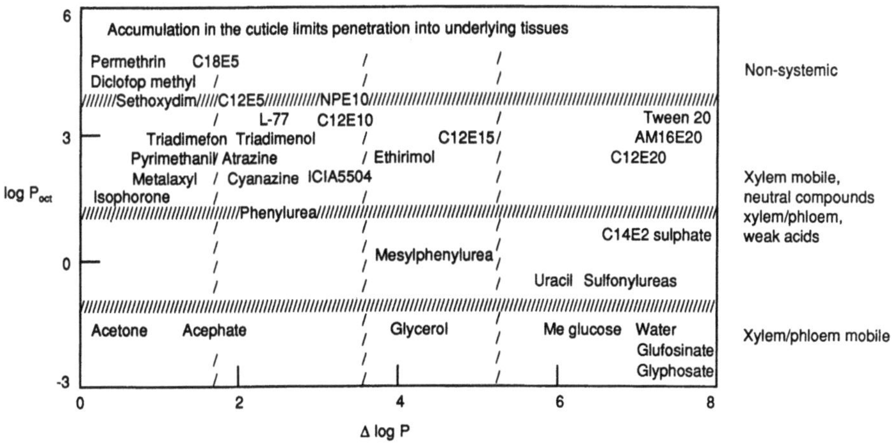

Fig. 5. Estimated log P_{oct} and Δ log P for active ingredients and adjuvants

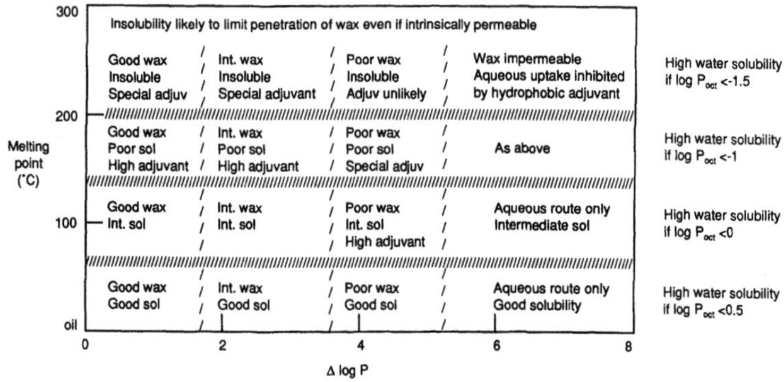

Fig. 6. Effect of melting point and Δ log *P* on inherent wax permeability and on solvent solubility

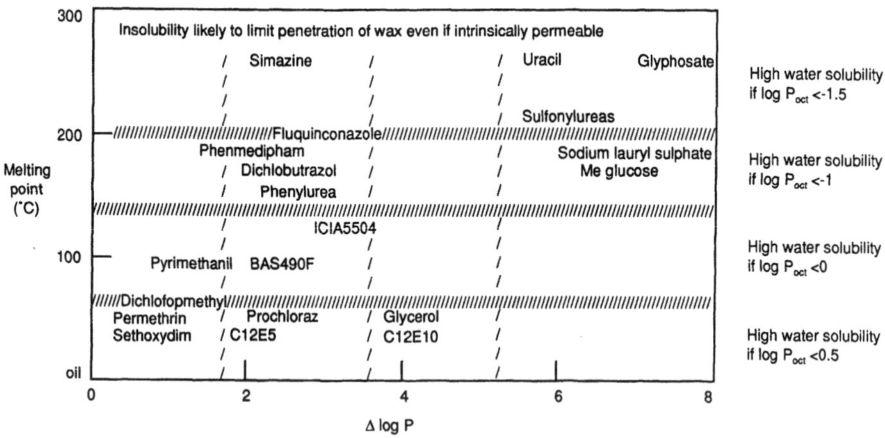

Fig. 7. Melting point and Δ log *P* of active ingredients and adjuvants

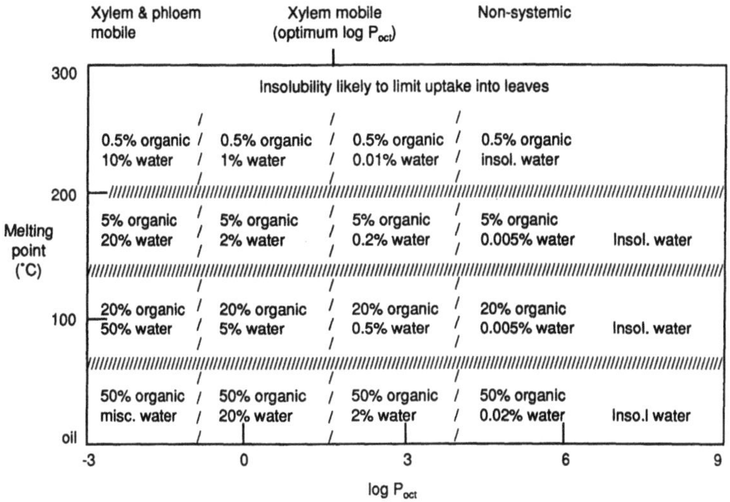

Fig. 8. Effect of melting point and log P_{oct} on systemic behaviour and on solubility (%w/v)

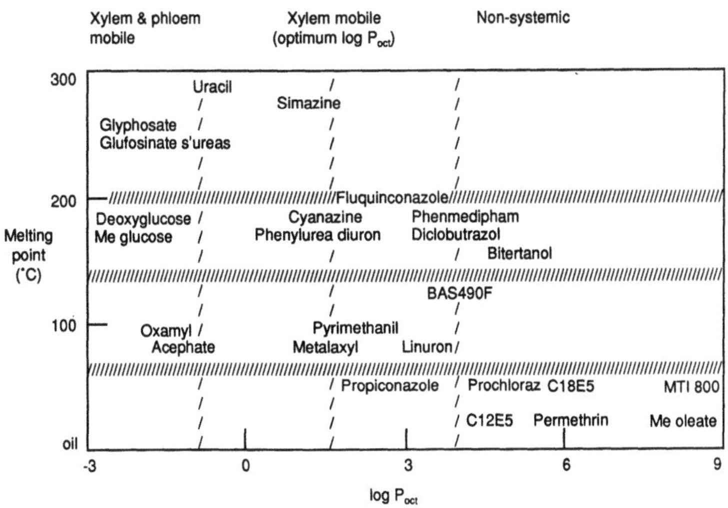

Fig. 9. Melting point and log P_{oct} of active ingredients and adjuvants

work involves compounds chosen because they have biological activity on whole organisms and must therefore penetrate to some extent and, thus, eliminates combinations unfavourable for uptake, and (b) some properties are associated. For example, large $\Delta \log P$ values arise from functionalities, such as amides and ureas, and these also have high melting points because they form stable intermolecular hydrogen bonds. The 1-methoxy-1-methylureas, such as linuron, form intramolecular hydrogen bonds, have smaller $\Delta \log P$ values and lower melting points than the related 1,1-dimethylureas; they are thus much more effective as post-emergence herbicides.

The ideas, outlined in Figs. 4–9, offer an explanation of much of the work on foliar uptake. It is unlikely that regression analysis can satisfactorily account for the interplay of properties and two possible modes of entry. Any number of correlations are possible depending on the compounds investigated. The effect of $\log P_{oct}$ reported by Chamberlain et al. (1987) probably arises because the compounds investigated varied only in $\log P_{oct}$ while melting point and $\log P$ were relatively constant.

An interesting feature observed when plotting adjuvants and active ingredients together is that a number of adjuvants have some properties that match those of the materials for which they are the most effective at enhancing foliar penetration. Examples in Fig. 5 are glufosinate and the C14E2 sulphate (Köcher and Kocur 1993), permethrin and short-chain ethoxylates,and glyphosate and long-chain amine ethoxylates. Silwet L-77 looks singularly incompatible with glyphosate, and the literature (Gaskin and Stevens 1993) indicates that once the immediate effect of stomatal infiltration is over it is antagonistic to glyphosate penetration in the absence of more appropriate adjuvants.

The major weakness of this discussion is that the large differences between plant species have not been addressed. We would hope, however, that the discussion would assist in deciding which questions should be asked to explain these differences.

1.6 Conclusions

Our conclusion about adjuvants is that their main function is as solubilising agents, most effective when they have properties appropriate for the active ingredient. The focus in much adjuvant work is on the surface-active ethoxylates; this must be seen as an accident of history rather than a rational discovery and has inappropriately skewed much research because of the expectation that their effectiveness as adjuvants lies in their surface activity. However, very different materials, such as the fatty acid derivatives that are solvents without surface-active properties, are also effective with active ingredients having appropriate chemistry.

If asked to design a simple polar solvent, which is non-volatile when applied to plants at 1 kg/ha, the simplest starting point for the synthetic chemist is the linear C_{24} alcohol (EH = 29), C_{26} ketone (EH = 28) or C_{26} ester (EH = 28).

Starting with the alcohol, which costs ca. 300 DM per gram, more economic alternatives would be sought. Using the well-tried isosteric substitutions of NH or O for CH_2 and an aromatic ring for $CH_2CH_2CH_2$ (Thornber 1979) combined with simple chemistry using readily available intermediates, the chemist would quickly make the following changes:

$CH_3(CH_2)_{10}CH_2CH_2(CH_2)_{10}CH_2OH$	Lead for modification	
↓		
$CH_3(CH_2)_{10}CH_2OCH_2(CH_2)_9CH_2OH$	Replace central CH_2 by O	Cheap C_{12} alcohol, C_{11} expensive
↓		
$CH_3(CH_2)_{10}CH_2O(CH_2CH_2O)_3CH_2CH_2OH$	Replace further CH_2 by O	Cheap C_{12} alcohol, cheap ethylene oxide

The NH or aromatic isosteric substitutions would lead to the alkylamine and alkylphenol analogues. The chemist has invented solvents meeting the simplest requirements of an adjuvant; these solvents also happen to be the major non-ionic surfactants. In real life, these are readily available as cheap, agriculturally acceptable agents because of their surfactant properties at low concentrations. At high concentrations, this surface activity is a coincidence and not a rationale for their adjuvant effects.

Alternatively, starting from the ester:

$CH_3(CH_2)_{23}COOCH_3$	Lead for modification	Expensive acid, cheap alcohol
↓		
$CH_3(CH_2)_{16}COOCH_2(CH_2)_6CH_3$	Move functionality	Cheap acid, cheap alcohol, ester solid
↓		
$CH_3(CH_2)_7CH = CH(CH_2)_7COOCH_2(CH_2)_6CH_3$	Double bond lowers melting point	Cheap acid, cheap alcohol

The adjuvant properties of octyl oleate are described by Gauvrit (this volume). Rational design of adjuvants is here!

Acknowledgements. The helpful suggestions for improvement of the manuscript from colleagues at Schering Agrochemicals Limited, especially David Stock, are gratefully acknowledged.

References

Anderson NH, Girling J (1983) The uptake of surfactants into wheat. Pestic Sci 14:399–404

Arnold KA, Wildman AS, White RJ, Buss MW, Cline MN (1993) Improved early symptom development with a ready-to-use glyphosate formulation by addition of fatty acid. Pestic Sci 38:270–271

Baker EA (1982) Chemistry and morphology of plant epicuticular waxes. In: Cutler DF, Alvin KL, Price CE (eds) The plant cuticle. Academic, London, pp 139–165 (Linnean Society symposium series, no 10)

Barak E, Dinoor A, Jacoby B (1981) Adsorption of systemic fungicides and a herbicide by some components of plant tissue in relation to some physicochemical properties of the pesticide. Pestic Sci 14:213–219

Baker EA, Hayes AL, Butler RC (1992) Physicochemical properties of agrochemicals: their effects on foliar penetration. Pestic Sci 34:167–182

Briggs GG, Bromilow RH, Evans AA (1982) Relationships between lipophilicity and root uptake and translocation on non-ionised chemicals by barley. Pestic Sci 13:495–504

Briggs GG, Bromilow RH, Evans AA, Williams M (1983) Relationships between lipophilicity and the distribution of non-ionised chemicals in barley shoots following uptake by the roots. Pestic Sci 14:492–500

Briggs GG, Rigitano RLO, Bromilow RH (1987) Physicochemical factors affecting the uptake by roots and translocation to shoots of weak acids in barley. Pestic Sci 19:101–112

Bromilow RH, Chamberlain K (1991) Pathways and mechanisms of transport of herbicides in plants. In: Kirkwood RC (ed) Target sites for herbicide action. Plenum, New York, pp 245–284

Bromilow RH, Chamberlain K, Evans AA (1990) Physicochemical aspects of phloem translocation of herbicides. Weed Sci 38:305–314

Chamberlain K, Briggs GG, Bromilow RH, Evans AA, Chen QF (1987) The influence of physicochemical properties of pesticides on uptake and translocation following foliar application. Aspects Appl Biol 14: 293–304

Gaskin RE, Stevens PJG (1993) Antagonism of foliar uptake of glyphosate into grasses by organosilicone surfactants. I. Effects of plant species, formulation, concentrations and timing of application. Pestic Sci 38:185–192

Hartley GS, Graham-Bryce IJ (1980) Physical principles of pesticide behaviour. Academic, London

Higuchi T (1977) Prodrug, molecular structure and percutaneous delivery. In: Roche EB (ed) Design of biopharmaceutical properties through prodrugs and analogs. American Pharmaceutical Association, Washington, pp 409–421

Köcher H, Kocur J (1993) Influence of wetting agents on the foliar uptake and herbicidal activity of glufosinate. Pestic Sci 37:155–158

Krenek MR, Reed KG, King DN (1987) Factors affecting the phytotoxicity of solvents used in pesticide formulations. In: Greenhalgh R, Roberts TR (eds) Proceedings 6th international congress of pesticide chemistry. Blackwell, Oxford, pp 287–290

Leahy DE, Morris JJ, Taylor PJ, Wait AR (1992) Model solvents for QSAR. II. Fragment values (f-values) for the "critical quartet". J Chem Soc Perkin Transact 2:723–731

McCall PJ (1988) Quantitative modelling of the effects of physical and chemical parameters on the foliar penetration of pesticides and its potential for predicting field behaviour. Aspects Appl Biol 21:185–201

Nalewaja JD, Matysiak R (1993) Influence of diammonium sulphate and other salts on glyphosate phytotoxicity. Pestic Sci 38:77–84

Schönherr J, Riederer M, Schreiber L, Bauer H (1990) Foliar uptake of pesticides and its activation by adjuvants: theories and methods for optimisation. In: Frehse H (ed) Proceedings 7th international congress of pesticide chemistry. VCH, Weinheim, pp 237–253

Silcox D, Holloway PJ (1986) Techniques suitable for the determination of foliar uptake, movement and metabolism of nonionic surfactants. Aspects Appl Biol 11:1–11

Stevens PJG, Bukovac MJ (1987) Studies on octylphenol surfactants. I. Effects of oxyethylene content on properties of potential relevance to foliar absorption. Pestic Sci 20:19–35

Stock D, Holloway PJ (1993) Possible mechanisms for surfactant-induced foliar uptake of agrochemicals. Pestic Sci 38:165–177

Stock D, Holloway PJ, Grayson BT, Whitehouse P (1993) Development of a predictive uptake model to rationalise selection of polyoxyethylene surfactant adjuvants for foliage-applied agrochemicals. Pestic Sci 37:233–245

Tayar NE, Ruey-Shiuan T, Testa B, Carrupt P-A, Hansch C, Leo A (1991) Percutaneous penetration of drugs: a quantitative structure-permeability relationship study. J Pharm Sci 80:744–749

Thornber CW (1979) Isosterism and molecular modification in drug design. Chem Soc Rev 8:563–580

Uchida M (1980) Affinity and mobility of fungicidal dialkyl dithiolanylidene-malonates in rice plants. Pestic Biochem Physiol 14:249–255

Urvoy C, Pollacsek M, Gauvrit C (1992) Seed oils as additives: penetration of triolein, methyl oleate and diclofop-methyl in maize leaves. Weed Res 32:375–383

Worthing CR, Hance RH (1991) The pesticide manual, 9th edn. British Crop Protection Council, Farnham

Young RC, Mitchell RC, Brown TH, Ganellin CR, Griffiths G, Jones M, Rana KK, Saunders D, Smith IR, Sore NE, Wilks TJ (1988) Development of a new physicochemical model for brain penetration and its application to the design of centrally acting H2-receptor histamine antagonists. J Med Chem 31:656–671

Zabkiewicz JA, Stevens PJG, Forster WA, Steele KD (1993) Foliar uptake of organosilicone surfactant oligomers into bean leaf in the presence and absence of glyphosate. Pestic Sci 38:135–143

2 Foliar Penetration of Agrochemicals: Limitations Imposed by Formulation Type and Some Interactions with Adjuvants

D. Stock and L. E. Davies

2.1 Introduction

The influence of formulation and adjuvants on the biological activity of a pesticidal active ingredient (AI) is a complex interactive phenomenon, with many factors possibly influencing the final response observed.

These factors include the many steps and processes between the spray tank and the biochemical site of action in the target organism, be it plant, pathogen or insect. However, some of them are beyond the control of the formulation chemist. For example, the movement and metabolism of an active ingredient are determined by the physicochemical properties of the compound and the physiological/biochemical characteristics of the target organism, respectively. Changes to the physicochemical properties of the active ingredient by synthesis of analogues may lessen the desired biological impact. The physicochemical properties which the formulation chemist must work with are therefore fixed by biological screening data, unless the AI is suitable for the synthesis of a salt. The resultant physicochemical properties of an active compound may not necessarily be optimal for easy delivery to the biochemical site of action (Stock 1991). Optimising delivery of the active ingredient within the constraints imposed by its physicochemical properties depends upon the skill of the formulation chemist married with the experience and experimental expertise of the biologist.

The purpose of this contribution is to show how the physicochemical properties of an AI influence the type of formulation which can be produced. In addition, the biological merits of different types of formulation are discussed. A case study using a model fungicide is used to illustrate the dilemmas encountered during the biological optimisation of a new compound, a process which can be made more difficult at an early stage by the lack of knowledge of the compounds mode and biochemical site of action. One factor in particular which can significantly influence activity, foliar penetration, is explored in detail. It is shown that different types of formulation show very different penetration responses, whilst their response to the inclusion of an adjuvant can vary greatly.

Whilst the physicochemical properties of an active ingredient limit the type of formulations which can be produced, and the activity of these variants is further used to narrow the optimisation process, other factors also influence the selection process during commercial development. It must be possible to produce the formulation reliably on a commercial scale and at an acceptable cost whilst complying with ever more restrictive registration criteria. As a constant background to this logical series of prioritised choices there is, of course, the requirement to obtain a large enough percentage of the world market to allow a venture to proceed. Superimposed on this final economic judgement are many subjective factors such as market fashion in particular countries and the perceived attractiveness of a novel product. In such a commercial minefield it is indeed amazing that new AIs and formulations thereof are reaching the consumer, albeit at an ever decreasing, and more finely judged, rate!

2.2 Formulation Design: Influence of Physicochemical Factors

It quickly becomes apparent that AIs do not always have appropriate physicochemical properties to be formulated in their most biologically active form. The choice of formulation from a physicochemical viewpoint is, in many cases, the most unsuitable for expressing activity. For example, a high melting point systemic herbicide which requires efficient penetration for maximal activity is in its least active form if applied to the leaf surface as suspended particles. Thus, formulation development often results in compromise.

The most important factors which determine the feasibility of a formulation type are: physical form of the AI, melting point, solubility characteristics in aqueous and organic media and chemical stability. The importance of such characteristics is well established (Scher 1984). Figure 1 summarises the main choices available during a formulation exercise. If this series of prioritised choices is related to the requirements for optimal biological activity it can be appreciated that conflict often occurs. This has led to the development of more sophisticated types of formulation such as suspoemulsions and oil-based suspension concentrates (SCs) in addition to the evaluation of many surface active materials which may have dual activity by imparting biological enhancement in addition to required interfacial properties.

2.3 Biological Properties of Formulations

The relative merits of different types of traditional agrochemical formulation are based on many criteria, not necessarily biological. For exam-

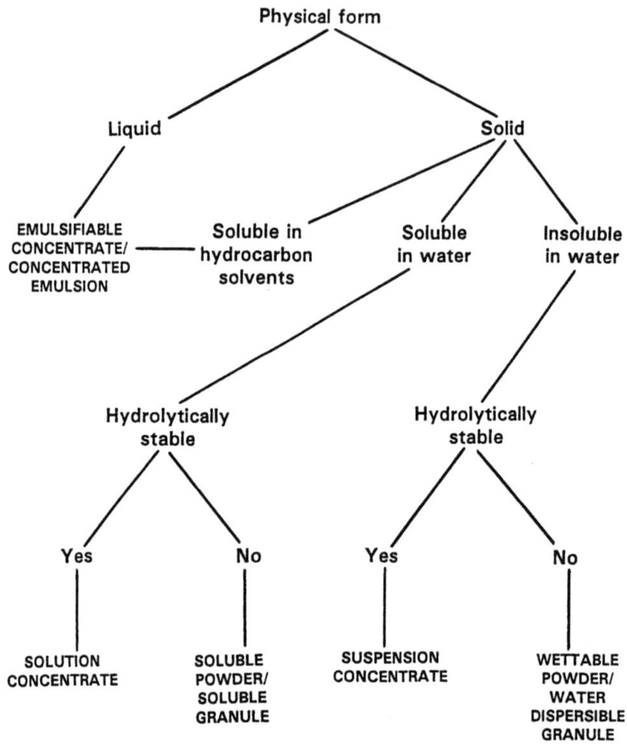

Fig. 1. Influence of the physicochemical properties of pesticidal active ingredients on formulation opportunities

ple, safety for transport and operator handling are of increasing importance. These primarily non-biological factors are discussed in Sect. 2.5.

In deciding the optimal type of formulation for biological activity many factors must be considered, some of which are only relevant to a particular crop situation. These include the persistence of activity over an acceptable time frame in addition to an assessment of undesirable side effects such as phytotoxicity. The requirement for persistence depends upon the crop and type of application regime routinely used. The requirement for persistence of (safe) activity is most frequently associated with fungicide and insecticide/acaricide applications whilst reliable first-hit action is the predominant success criterion for herbicides.

2.3.1 Biological Activity of Formulations

Much of the current body of information regarding the biological activity of different types of formulation is based on individual experience, specifically by workers in the agrochemical industry, together with myth and personal preference from the end user. Rarely is there any open literature regarding the biological activity of one AI presented as different types of formulation. Often it is not possible to produce widely differing types of formulation due to the restrictions imposed by the physicochemical properties of the compound (Sect. 2.2). Inevitably, much of the judgement about formulations is therefore subjective and based on comparisons of different AIs used for similar crop applications.

The current commercial judgement of the major types of formulation in common usage is that solvent-based products, such as emulsifiable concentrates (ECs), are generally the most active. They are also the least favoured environmentally and for operator exposure (in particular dermal penetration) and safe transportation. Production of low solvent products such as concentrated emulsions (EWs), in which the external phase is aqueous, is a compromise. However, the complete avoidance of solvent in formulations may increasingly necessitate the use of biologically less suitable types of formulation which, in turn, could lead to a requirement for an increased dose of AI.

Of the major non-solvent-based formulations commonly used, those with a particulate content, such as wettable powders (WPs) and aqueous SCs are often viewed as being less active, particularly for herbicides which require foliar penetration to reach their site of action. It is with such "less active" types of formulation that inclusion of a tank-mix adjuvant is sometimes required in order to boost biological activity. Obviously, development of an all-in-one product is judged to be optimal. However, tank mixing of adjuvants to boost activity is necessary when the most effective adjuvants are not compatible within a single formulation. For example, inclusion of a high proportion of an activator surfactant in an SC formulation may lead to excessive crystal growth of particles of AI. If oil adjuvants are shown to produce good biological enhancement of a particular AI, the development of an oil SC or a suspoemulsion may be pursued. The former relies upon minimal solubility of the AI in the carrier, whilst the latter may be pursued if this is not the case. These formulations are currently of greater interest, although formidable technical problems are often encountered in the production of a chemically and physically stable product with satisfactory storage and handling properties.

For necessarily dry formulations, such as WPs, water dispersible granules and (at the extreme limit for low rate AIs) tablet formulations, the inclusion of a significant quantity of activator within the product is difficult to achieve, particularly if the required activator is liquid. Avoidance of tank-mixing with such products may be achieved by the use of novel packaging techniques, such as partitioning the incompatible components with water soluble packaging, for example, the use of bag-in-bag systems. Liquid components can be stabilised in such systems, and the use of gelled products can provide suitable handling properties.

In acknowledging that different types of formulation are perceived as more or less active depending upon the type of pesticide and crop area, answers must be sought to explain the various biological advantages and failings. A clear understanding of the major determinant factors with respect to biological activity should allow more rational commercial development of an AI. Differences in biological activity may be attributed to a combination of the factors discussed in the following three sections.

2.3.1.1 Effect of Formulation on Foliar Penetration

Little open literature is available regarding the foliar penetration of different types of agrochemical formulation. Most published work refers to the penetration properties of active ingredients in idealised systems, such as aqueous acetone for whole plant applications (Stock et al. 1993; Urvoy and Gauvrit 1991), aqueous systems in isolated cuticle studies (Schönherr and Riederer 1988) or detached, immersed leaves (Schreiber and Schönherr 1992). Limited uptake studies have, however, been made using the commercial product (Holloway et al. 1992; Wirth et al. 1991). Such studies are most accurately conducted if a radiolabelled formulation is available, although other analytical methods such as HPLC may be used if the compound has a good chromophore and/or is used at a sufficiently high rate. Even in reports which involve investigation of the formulated product, parallel comparison of different types of formulation of the same AI is rarely addressed. It is usually assumed that, if an AI can be delivered to the leaf surface solubilised in an organic phase, the potential for foliar penetration is increased. However, the manufacture of such a product is not always possible due to the physicochemical limitations imposed by the AI (Fig. 1).

The lowest potential for foliar penetration would seem to be from a particulate formulation, particularly if the AI is highly crystalline with a high lattice energy. In such systems the AI is delivered to the leaf surface as discrete particles with mass median diameters (MMDs) typically in the range 1–5 m. However, even when delivered in such an apparently unsuitable physical form, foliar penetration of the AI can be significant in the presence of suitable "inert" ingredients, such as surfactants. Indeed, Holloway et al. (1992) showed significant penetration of ethirimol and diclobutrazol into wheat leaves from SC formulations when suitable alcohol ethoxylates at sufficient concentrations were added. It is often assumed that such penetration is attributable to solubilisation of the AI in surfactants, which acts as a transfer phase to the leaf surface. The solubilisation capacity of surfactants for AIs can be significant (Temple and Hilton 1963; Nassetta et al. 1991). Nevertheless, the study of Holloway et al. (1992) showed that, under the experimental conditions used, the uptake activating potential of the non-ionic surfactants used could not be attributed to their capacity to solubilise the AIs. The interaction between the AI and surfactants as they penetrate through the leaf surface has also been shown to be of great significance (Stock et al. 1992). Studies with a range of model compounds, including glyphosate, difenzoquat, methylglucose and an experimental tetrazole herbicide, have shown that both interactive and non-interactive mechanisms of surfactant-induced uptake occur. The former mechanism is predominant with lipophilic compounds which generally reduced the penetration of lower alcohol ethoxylate surfactants during activation of foliar uptake. Irrespective of the effects of surfactants on the penetration of an AI from an SC formulation, the carrying capacity of SCs for surfactant can be very limited, depending upon the physicochemical properties of the AI and the biologically beneficial surfactant. Use of a tank-mix surfactant may therefore be the only alternative in such a situation.

A possible method for enhancing activity of particulate formulations, without the requirement for tank-mixing with adjuvants, is the development of an oil-SC, the so-called oily flowable or a suspoemulsion containing an emulsified oil. The activity, if not penetration, potential of such formulations can be great. Work by Nalewaja's group

(Nalewaja 1986; Nalewaja et al. 1990) has shown the biological activation potential of oils and their derivatives with a selection of herbicides. Oils tested include plant oils and the corresponding fatty acid ester derivatives as well as mineral oils. A more recent study by this group using the herbicide quizalofop-ethyl (Manthey et al. 1992) has shown that the enhanced activity with lipid additives is usually associated with enhanced foliar penetration. Similarly, Gauvrit and coworkers have found that oil additives increase the penetration of diclofop-methyl into grass species (Gauvrit and Dufour 1990; Schott et al. 1991). It may therefore be surmised that such activating potential could be built into an oil-SC or suspoemulsion rather than by using a tank-mix crop oil or crop oil concentrate. Depending upon the physicochemical properties of the AI, development of an oil-based EC may also be a possibility.

A soluble concentrate, SC formulation, containing the AI dissolved in water is a type of formulation from which foliar penetration can be significant. This type of formulation is obviously limited to AIs which possess sufficient water solubility and hydrolytic stability, such as 2,4-D and glyphosate. However, substantial solubility is rare in the modern generation of pesticidal active ingredients because it often accompanied by unacceptable leaching properties which would hinder registration. Nevertheless, water soluble compounds can show very significant foliar penetration even into water repellant leaf surfaces, particularly in the presence of suitable surfactants/adjuvants (Stock et al. 1993). In addition, formulants with humectant properties can significantly enhance the penetration of such compounds (Cook et al. 1977; Ben Tal 1987). The prevention of recrystallisation during droplet drying is often invoked as a reason for efficient foliar penetration (MacIsaac et al. 1991; Nalewaja et al. 1992). The importance of the physical form of deposits is discussed in detail by Falk (this volume). A unique property of water-soluble compounds would seem to be the potential for foliar entry via open stomata when a suitable adjuvant is used. Organosilicone surfactants are at present the only agents capable of specifically promoting this route of entry (Field and Bishop 1988; Stevens et al. 1991). Organosilicones are generally recommended for tank-mixing, although inclusion within a formulation is possible (Dayawon et al. 1989). However, the importance of the stomatal route of entry in the presence of silicone-based surfactants has been recently challenged (Roggenbuck et al. 1993).

2.3.1.2 Effect of Formulation on Foliar Retention of Spray Solutions

Foliar retention of spray applications, particularly in relation to activity and selectivity, has been extensively covered in the literature over several decades. Indeed, differential retention by crop and weed species has been invoked as an important mechanism of selectivity for some herbicide applications (Davies et al. 1967; Blackman et al. 1958; Holly 1976; Hibbitt 1969). Much of the early work in this area has centred on differences in leaf surface structure and plant canopy structure rather than the physical properties of commercial formulations. In addition, any treatment of the mechanism of retention has generally been restricted to a theoretical treatment based on information available at the time. Predictive tailoring of formulations to increase or decrease foliar retention on particular crops has therefore not been pursued with any great success. Indeed, the factors which control foliar retention are not precisely understood. Whilst leaf surface factors obviously play a major role (Watanabi and Yamaguchi 1991), the physical properties of the spray solution which can modify retention need to be fully understood.

The dynamic surface tension of simple spray solutions has in recent years been shown to be a major factor in determining their foliar retention (Anderson and Hall 1989; de Ruiter et al. 1990; Grayson et al. 1991). However, whilst dynamic surface tension of liquids containing surfactant is a good predictor of retention, it is not for solutions of polyvinyl alcohols (Wirth et al. 1991). Most of these studies have involved the use of idealised solutions containing one surfactant, rather than complex mixtures of components characterised by realistic agrochemical formulations. If one considers the typical composition of the major types of agrochemical formulation, specifically the type and content of surface active agents, the potential for foliar retention is likely to vary greatly. Indeed, it is often the observation that EC formulations are qualitatively retained more effectively than WPs. Whilst WPs can contain fairly high proportions of surface active agents to promote wetting and dispersion of particles upon addition to the tank, it must be remembered that the AI content of WPs is often high and this can result in a low AI: surfactant ratio. In addition, the wetting/dispersing agents for the AI are generally not the best materials for the wetting of foliage. Requirement for compatibility with a dry formulation complicates the choice of surfactant which can be used in a WP for spray modification and/or biological activation purposes. Limited quantities of liquid surfactant may be adsorbed onto silica for incorporation into a WP, but the use of silica can in itself cause application problems. By comparison, ECs can contain relatively large amounts of surfactant, which could explain the visually good retention of this type of formulation. Grayson et al. (1993) found good retention of ECs compared with other formulation types in their track-sprayer investigations. In addition, studies by Wirth et al. (1991), using experimental formulations of tebuconazole, showed that EC and EW formulations gave better retention than WPs, in particular, those with a low surfactant content. The angle of inclination of the leaf used in these studies had little effect on overall retention.

Whilst studies under controlled conditions have shown the relevance of dynamic surface tension to retention phenomena of surfactant-containing solutions, reduction of this parameter is not a guarantee of good retention; much depends upon the "wettability" of the leaf surface. For example, Holloway (1993) showed that the organosilicone Silwet L77 which is known to reduce dynamic surface tension, caused a significant reduction in spray retention on the leaves of sugar beet compared with conventional organic surfactants and even water! These findings confirm the early observation of Furmidge (1962) that requirements for retention and coverage can be conflicting. If the coverage of a leaf surface is effective (characterised by a low advancing contact angle), mean retention per unit of area can be considerably reduced. The importance of foliar coverage versus retention depends upon the mode and site of action of the AI.

2.3.1.3 *Effect of Formulation on Foliar Distribution of Pesticides*

Owing to the widely varying properties of the main types of agrochemical formulation, there is a large potential for variation in the nature of foliar deposits and their distribution. Like foliar retention, the physical factors which determine foliar distribution are not fully understood as it involves a complex interaction governed by the topography and chemical nature of the leaf surface and the physical properties of the diluted spray formulation. In addition, when considering foliar distribution one must consider the area of spread and distribution of spray droplet in addition to the distribution of AI within the apparent contact/distribution area.

A number of physical parameters may be considered when attempting to understand the foliar distribution and spread of spray droplets. It is well established that there is no simple relationship between foliar spread and the equilibrium surface tension of the spray solution. Indeed, studies by Abbott et al. (1990) using a wide selection of pure organic liquids and adjuvants, many of which were emulsifiable oils, showed that quantifying spread in relation to a particular physical parameter such as equilibrium surface tension, contact angle and critical micelle concentration (CMC) was inconclusive. The possibility of pursuing such an approach with a complex mixture such as an agrochemical formulation would seem daunting.

For particulate formulations it is proposed that, as the droplet spreads after initial impaction, particles at the leading edge of the droplet lead to adhesion and contact angle hysteresis during retraction. A larger area of contact than that from a true solution can therefore result. However, further foliar spread is usually expected and desired after this initial impaction process. For aqueous based applications containing organosilicone surfactant the amount of foliar spread can be spectacular. This spread has been linked to the alignment and adsorption of the silicone surfactant at the droplet-leaf interface (Murphy et al. 1991). However, whilst extensive spread can occur giving potential for uniform film coverage of the leaf surface, a reduction in mean overall deposition can also result (Holloway 1993). It must be remembered that maximum retention occurs at incipient runoff (Furmidge 1962).

The presence of emulsified solvents/oils, characteristic of ECs, can significantly enhance the potential for foliar spread, particularly on species which posses crystalline wax, such as many grass species. McWhorter and Barrentine (1988) found dramatic foliar spread of oils on Johnson grass. Whilst this spread was greatest with entirely oil-based applications, spread with emulsified oil was also significant. A similar finding was also noted by Baker et al. (1983) in their investigation of oil-based solutions applied to a selection of plant species. They showed that spread was limited on species such as sugar beet and lemon, which have a smooth leaf surface, whilst spread was significant on species which possessed crystalline wax or a pronounced leaf microstructure (trichomes, secretory glands, etc).

The need for good distribution, both at a macro- and amicro-level, is dependent upon the physicochemical properties of the AI and its site of biochemical action. For a systemic herbicide, uniform distribution need not be essential. Even for "contact" herbicides, total coverage may not be required provided that sufficient tissues are exposed to the effects of the penetrating herbicide. However, for fungicides the problem is more difficult. If a fungicide is non-systemic, good foliar coverage would seem essential. For a particulate formulation, this requirement is especially great because it provides only point sources of activity within the deposit. Leaf areas not covered by deposit may be unprotected depending upon the sphere of influence of the particles of AI and their distribution. This sphere of influence could be particularly low for a high melting point AI with no vapour activity. Formulants in the deposit could conceivably increase bioavailabilty of the AI to pathogens depending upon the route of entry of the AI to its target site of action. Very little evidence is available regarding the mechanistic influence of formulation on fungicide activity, although there is no doubt that formulation and adjuvants can significantly influence activity, persistency and stability of fungicides (Backman 1978).

In considering the deposit area of a spray application it must be accepted that coverage of the AI is not uniform. Often the AI is deposited as an annulus at the perimeter of the evaporating droplet (Baker et al. 1983). In studies with a wide range of AIs dissolved in methanol-water, Baker et al. (1992) found that the inclusion of a nonylphenol 8EO surfactant increased the incidence of annulus formation. In addition, the formation of annuli varied according to leaf surface, although a statistically significant relationship could not be shown. The area of effective coverage may therefore be relatively small. ECs often give a qualitatively better coverage of AI within the apparent contact area.

Distribution with respect to the leaf surface/canopy is an additional consideration when attempting to optimise a formulation. Coupland et al. (1978) showed that penetration of difenzoquat and benzoylprop-ethyl into the leaf-base of wild oat enhanced biological activity, whilst Cooke et al. (1989, 1993) have shown that penetration of prochloraz down the inner leaf sheath of wheat enhances control of eyespot. It is therefore essential when designing a formulation that the coverage of the target be a major consideration. During the early stages of development of a new compound this problem can be worsened by the lack of knowledge regarding the mode of action of the AI. A certain amount of trial and error may therefore be unavoidable during the primary stages of investigation of a compound.

2.4 Influence of Formulation, Adjuvants and Their Interaction on the Foliar Penetration of SN 612352 into Wheat and Vine Leaves

The physicochemical constraints on formulation opportunities are discussed in Sect. 2.2. In addition, the perceived biological properties of the major types of agrochemical formulation have been outlined. In this section a case study, using a model fungicide SN 612352, is used to demonstrate the effects of producing two formulation variants of an AI.

It is with the physicochemical properties of SN 612352 (Table 1) that a formulation chemist is first confronted before making a decision about which type(s) of formulation can or should be produced. The properties of this compound readily allowed the development of experimental EC, SC and WP formulations. In the following sections the foliar penetration properties of the EC and SC are investigated in detail. Of course, in the early stages of evaluation of a new fungicide the desired location of the molecule may not be known. A fungicidal AI may be required on and/or in the plant, or in a particular physical form, for maximum activity against a pathogen.

Table 1. Physicochemical properties of SN 612352

Property	Value
Solubility, water	0.2 mg/l
Solubility, Solvesso 200	50–100 g/l
Log P^{a}	4.0
Melting point (°C)	110

[a]Log octanol-water partition coefficient, determined by HPLC.

2.4.1 Formulation of SN 612352

In order to study the penetration properties of the different formulations conveniently and accurately, ^{14}C-labelled SN 612352 was synthesized. Using the labelled compound, radiolabelled versions of the 5 EC and 10 SC formulations (Tables 2 and 3, respectively) were produced. Using appropriate quantities of labelled and unlabelled compound an exact copy of the 5 EC could be produced. However, the milling procedures for SC formulations could not be reproduced accurately for the radiolabelled compound, due to the small quantities necessarily involved. A prediluted SC was, therefore, produced by milling 10 mg AI in a Mixer Mill with appropriate quantities of the other formulants. This procedure resulted in an MMD slightly greater than that of the unlabelled formula-

Table 2. Composition of SN 612352 5 EC

Component	Amount (g/l)
SN 612352	50
Emulsifiers	100
N-Methyl pyrrolidone	50
Solvesso 200	790

Table 3. Composition of SN 612352 10 SC

Component	Amount (g/kg)
SN 612352	100
Antifoam	0.5
Dispersants	50
Veegum	7.5
Propylene glycol	105
Biocide	1.5
Kelzan	1.0
Water	734.5

tion, for which an MMD of 1.03 m was achieved (measured using a Malvern MasterSizer).

2.4.2 Effect of Formulation and Adjuvants on Foliar Penetration of SN 612352 into Vine and Wheat Leaves

To determine the influence of adjuvants on foliar penetration when used in combination with the EC and SC formulations, preliminary investigations were first conducted to identify adjuvants/surfactants likely to have beneficial effects on the foliar penetration of SN 612352.

Extensive work on conventional alcohol ethoxylates (Stock et al. 1993) has shown that lower ethoxylates have beneficial effects on the uptake enhancement of high log *P* lipophilic compounds, such as SN 612352. An initial investigation was therefore conducted to investigate the influence of ethoxylation of a selection of alcohol ethoxylates on the penetration of SN 612352. In all preliminary studies an aqueous acetone (1:1) formulation was used. SN 612352 was applied as a 1 g/l solution according to the procedure of Stock et al. (1993), resulting in an application equivalent to 500 g/ha in 200 l. The surfactants were used at 2.5 g/l. Initial data (not shown) indicated very minimal penetration of SN 612352 in the absence of surfactant, whilst the lower ethoxylate Genapol C050 (coconut fatty alcohol 5EO) had the greatest beneficial effect on foliar penetration.

In addition to identifying a conventional organic surfactant for further investigation with the EC and SC formulations, a range of oil-based adjuvants were also investigated. A similar experimental procedure to that used with the conventional surfactants was used, except a higher acetone content (85%) was required in order to obtain a satisfactory solution/dispersion. The Atplus 411F, 412 and 413 adjuvants based on mineral oil, rape oil and cotton seed oil, respectively, were used for investigation. Each adjuvant was used at 2.5 and 0.5% g/l. Whilst none of the adjuvants was, quantitatively, as effective at inducing penetration as Genapol C050, Atplus 411F was identified as the most effective oil based adjuvant. It was therefore used alongside Genapol C050 for investigation of uptake activation properties with the EC and SC formulations of SN 612352.

Table 4. Surface recovery (percentage) of radiolabel from ^{14}C-SN 612352 (1 g/l) from 5 EC and 10 SC formulations applied to wheat and vine in the absence and presence of two adjuvants

Treatment (g/l)	Vine		Wheat	
	1 Day	5 Days	1 Day	5 Days
AI (1)	100.6	92.7	107.8	102.9
EC (1)	82.1	78.6	75.4	54.0
SC (1)	91.8	88.6	86.9	83.7
AI (1) + Genapol C050 (2.5)	75.2	68.5	75.1	60.0
EC (1) + Genapol C050 (2.5)	71.8	67.2	33.6	32.9
SC (1) + Genapol C050 (2.5)	75.1	72.7	72.0	88.9
AI (1) + Aplus 411F (5)	83.6	69.8	69.0	54.9
EC (1) + Atplus 411F (5)	85.8	80.6	63.3	50.0
SC (1) + Atplus 411F (5)	88.5	77.3	83.6	84.7
LSD ($p = 0.05$)	10.5		8.0	

AI, Unformulated control.

Investigation of the influence of formulation and its interaction with adjuvants was conducted using technical AI dissolved in aqueous acetone as the control treatment. The AI was applied at 1 g/l, whilst Atplus 411F and Genapol C050 were used at 5 and 2.5 g/l, respectively. Uptake investigations were performed using wheat and vine as target species. All applications were made as ten 0.2 l droplets to the treatment leaf using a microapplicator (Stock et al. 1993). Uptake was assessed at 1 and 5 days after foliar application using the cellulose acetate film-stripping method (Silcox and Holloway 1986) to remove residual surface deposits for liquid scintillation counting (LSC).

The surface recovery data (Table 4), showed considerable differences between the formulations both in the presence and absence of adjuvant. In the absence of an adjuvant, penetration of radiolabel from ^{14}C-SN 612352 into wheat was greatest from the EC formulation: 54% surface recovery at 5 days after application compared with 84% with the SC. Uptake in the absence of formulants was negligible. Foliar penetration into vine, in the absence of adjuvant, was very limited for both types of formulation.

The response to inclusion of an adjuvant/surfactant varied significantly according to formulation and plant species. Uptake activation into vine was always limited. It is probable that a higher adjuvant threshold concentration is required to induce penetration into vine. Previous studies by Stock et al. (1992) have shown that a distinct concentration threshold often must be exceeded before significant uptake activation occurs. However, in the present studies the EC alone was phytotoxic to vine, whilst both Genapol C050 (2.5 g/l) and Atplus 411F (5 g/l) caused the SC to be phytotoxic. Any increase in concentration of activators was thus not viable.

The uptake response of both formulations applied to wheat showed a significant variation in performance. Addition of Genapol C050 to the EC resulted in significant uptake activation by 1 day after foliar application, with 34% surface recovery compared with 75% in the absence of surfactant. The response by the SC treatment to the inclusion of Genapol C050 was minimal at both 1 and 5 days after application, with no evidence for uptake activation.

The response to the inclusion of Atplus 411F of the EC and SC applied to wheat was similar to that observed for the inclusion of Genapol C050, although less pronounced. Differences were slower to develop, with the greatest differences seen after 5 days. However, at this sampling interval inclusion of Atplus 411F had not increased the magnitude of penetration of AI from the EC, compared with the EC applied alone, whilst surface recovery of the SC was similar both in the presence and absence of Atplus 411F (ca. 84%).

2.4.3 Influence of EC Formulation Components on the Foliar Penetration of SN 612352

The experimental data described in the previous section show that there is a significant difference in uptake performance between the SC and EC formulations of SN 612352. A study was therefore conducted to determine why the EC induced greater uptake activation. Both the EC and SC formulations were applied to young vine plants as control treatments for two experimental compositions: (a) 10 SC + EC emulsifiers and (b)

Table 5. Effect of EC emulsifiers on the surface recovery (percentage) of radiolabel from ^{14}C-SN 612352 (1 g/l) from 5EC and 10SC formulations after foliar application to vine

Treatment (g/l)	Time	
	1 Day	5 Days
10SC (1)	96.3	91.5
10SC (1) + EC emulsifiers	97.0	90.7
AI (1) + EC emulsifiers	94.8	84.7
5EC (1)	54.1	31.1
LSD ($p = 0.05$)	10.9	

AI, Unformulated SN 612352

technical AI + EC emulsifiers. For both of these treatments, the EC emulsifiers were incorporated at the equivalent AI: emulsifier ratio as the parent EC formulation (Table 2). The treatment incorporating technical AI was made in 50% aqueous acetone.

The surface recovery data for the four treatments (Table 5) showed that addition of the EC formulants to either technical AI or the 10 SC formulation did not induce uptake comparable to that of the parent 5 EC. Indeed, there was no statistically significant increase in uptake caused by the addition of EC formulants to the SC or technical AI, for which a surface recovery of greater than 84% occurred. However, significant penetration of radiolabel into young vine plants occurred with the 5 EC, with ca. 50% surface recovery of radiolabel at 1 day after foliar application, decreasing to 30% at 5 days (Table 5). These findings provide strong circumstantial evidence for a requirement for the AI to be solubilised in the aromatic solvent/cosolvent in order to show maximal penetration. Clearly, delivery in an aqueous acetone system, in which the acetone rapidly volatilises away, does not result in comparable uptake. The reason for this difference is not clear. It is possible that solvent changes the partitioning into and mobility of the solubilised AI within the cuticle, analogous with such changes induced by penetrating surfactants (Riederer and Schönherr 1990).

2.4.4 Influence of Rate and Ratio of Surfactant: SN 612352 on Foliar Penetration from 5 EC Formulation into Vine and Wheat Leaves

Whilst it has been shown that the EC formulation induces greater uptake of SN 612352 than the particulate SC formulation, and that this can be further increased by the addition of an appropriate surfactant, the role of AI: surfactant ratio must also be considered. A study was performed to investigate this factor with the 5 EC formulation and Genapol C050. The EC was used at AI application rates of 0.25, 0.5 and 1 g/l (equivalent to 50, 100 and 200 g/ha, respectively), whilst the surfactant was applied at 0.5, 1 and 2.5 g/l. This resulted in a matrix of AI: surfactant ratios varying between 1:10 and 2:1 (Table 6).

The usual droplet application procedure (10 0.2 l droplets) was used to apply the treatments to both vine and wheat leaves. Sampling to determine the amount of uptake activation was performed at 1 and 5 days after foliar application. Most of the uptake activation occurred within the first day of foliar application, with a minimal further decrease in surface recovery by 5 days (full data not shown). A significant difference in response was noted between vine and wheat. For the treatments applied to vine (Fig. 2) containing AI at 0.25 and 0.5 g/l, an increase in Genapol C050 concentration from 0.5 to 2.5 g/l resulted in a decrease in surface recovery of radiolabel, i.e. increased foliar penetration. However, when the AI was applied at 1 g/l as the EC formulation, an increase in Genapol concentration resulted in a highly significant increase in surface recovery, i.e. decreased penetration. The reason for

Table 6. Matrix of SN 612352: Genapol C050 ratios investigated for foliar uptake properties

Rate of SN612352 (g/l)	Rate of Genapol C050 (g/l)		
	0.5	1.0	2.5
0.25	1:2	1:4	1:10
0.5	1:1	1:2	1:5
1.0	2:1	1:1	2:5

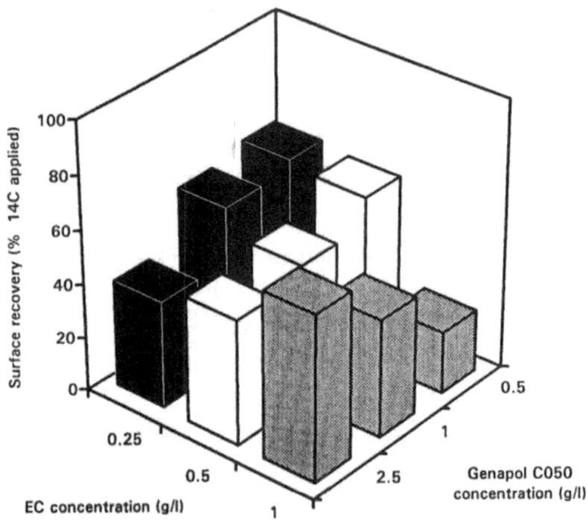

Fig. 2. Interaction between compound rate and surfactant rate on the surface recovery (%) of radiolablel from ^{14}C-SN 612352 5 days after foliar application to vine. LSD (p = 0.05) = 11.3

this difference is unknown. A similar phenomenon was noted by Gaskin and Holloway (1992) for the decreased foliar penetration of glyphosate into field bean in the presence of a high concentration of aliphatic amine surfactant. For the AI application rates of 0.25 and 5 g/l, the percentage surface recovery was not significantly different at the three Genapol C050 rates. A significantly greater mass penetration thus occurs at the 0.5 g/l rate. It is clear from the data that maintenance of a particular AI: surfactant ratio does not result in a consistent percentage uptake. For example, there is considerably (20%) less surface recovery of radiolabel for the treatment containing SN 612352 at 1 g/l and Genapol at 1 g/l (1:1 ratio), than for the equivalent treatment with both ingredients at 0.5 g/l. An important influence of surfactant rate and its interaction with compound rate can therefore be surmised. It is unlikely that any of the differential uptake responses according to composition could be attributable to phytotoxicity. All treatments, except that containing AI at

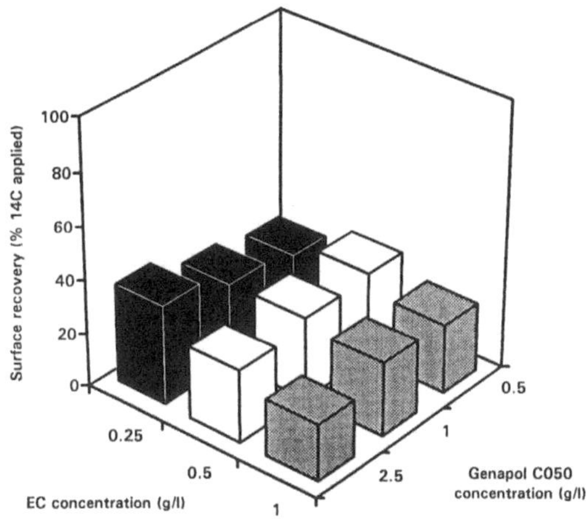

Fig. 3. Interaction between compound rate and surfactant rate on the surface recovery (percentage) of radiolablel from 14C-SN 612352 5 days after foliar application to wheat. LSD ($p = 0.05$) = 8.8

0.25 g/l with Genapol C050 at 0.5 g/l induced similar moderate to severe phytotoxic symptoms.

For the applications to wheat (Fig. 3), the differences in surface recovery were less pronounced than for vine. However, statistical conclusions could still be drawn. Unlike the situation with vine, the surface recovery for the AI at 0.25 g/l was greater when Genapol C050 was used at 2.5 g/l (38% surface recovery of radiolabel) than 0.5 g/l (27% surface recovery of radiolabel), i.e. increasing surfactant concentration had a negative effect on foliar penetration. However, for AI application rates of 0.5 and 1 g/l, changes in Genapol C050 concentration within the range 0.5–2.5 g/l had no significant influence on foliar penetration. As with vine, there was no consistent relationship between AI: surfactant ratio and uptake activation. The differences in results obtained according to plant species provide a good demonstration of the complex nature of formulation-surfactant-plant interactions. It is notable that the surfactant activator used in this investigation, Genapol C050, could be used as an emulsifier within a solvent-based formulation, rather than selecting it as an uptake-activator adjuvant. It may therefore be more appropriate to consider the findings as an example of formulation composition-plant species interaction, since changes in the emulsification system used in an EC formulation could be as important as the choice of adjuvant/surfactant for biological activation.

2.5 Further Constraints on Formulation Development

The present studies conducted with SN 612352 show that the interactions between formulation and adjuvants during foliar penetration, which is only one factor relevant to biological activity, are complex and are species dependent. This matrix of interactions is widened further if one considers the influence of environmental factors such as temperature, humidity and the potential for loss of AI by photodegredation and volatility. All of these factors can significantly influence the development of a product, in addition to the intrinsic physicochemical constraints. Other considerations can also be of great significance in determining the direction of development. Principal amongst these factors are the following: (a) *Economics*. The most important figure is cost per treatment. Increased raw material costs imposed by using activity enhancing formulations can be significant, but if this results in a significantly lower use rate per hectare, the economics may be favourable. Conversely, if a costly adjuvant increases the activity only slightly, development may be terminated. (b) *Processing*. The reproducibility, safety and economics of the proposed production process is an important consideration in the product development phase. The cost of processing of more complex types of formulation may be prohibitive in certain areas. (c) *Safety*. The product must be safe to both the spray operator and the environment. A formulation designed to increase penetration may improve activity greatly whilst increasing crop residue to unacceptable levels and/or increasing toxicity. (d) *Registration*. The suitability for registration of any new surface-active material which may be chosen for formulation/adjuvant purposes must be clarified before full development can take place. The actual choice of materials can be limited.

This list is not exhaustive, but highlights the variety of conflicting parameters of which the formulation chemist must be aware, and the reasons why compromise is often required during the development of a given formulation.

References

Abbott HA, van Dyke LP, Grobbelaar N (1990) Spreading of spray mixtures on leaf surfaces. I. Relative effectiveness of various physicochemical predictors. Pestic Sci 28:419–429

Anderson NH, Hall DJ (1989) The role of dynamic surface tension in the retention of surfactant sprays on pea plants. In: Chow PNP, Grant CA, Hinshalwood AM, Simundsson E (eds) Adjuvants and agrochemicals, vol 1. CRC, Boca Raton, pp 51–62

Backman PA (1978) Fungicide formulation: relationship to biological activity. Annu Rev Phytopathol 16:211–237

Baker EA, Hunt GM, Stevens PJG (1983) Studies of plant cuticle and spray droplet interactions: a fresh approach. Pestic Sci 14:645–658

Baker EA, Hayes AL, Butler RC (1992) Physicochemical properties of agrochemicals: their effect on foliar penetration. Pestic Sci 34:167–182

Ben-Tal Y (1987) Improving ethephons effect on olive fruit abscission by glycine. Hort Sci 22:869–871

Blackman GE, Bruce RS, Holly K (1958) Studies in the principles of phytotoxicity. V. Interrelationships between specific differences in spray retention and selective toxicity. J Exp Bot 9:175–205

Cook GT, Babiker AGT, Duncan HJ (1977) Penetration of bean leaves by amitrole as influenced by adjuvants and humidity. Pestic Sci 8:137–146

Cooke BK, Hislop EC, Jordan VWL, Western NM, Herrington PJ (1989) Redistribution of foliar surface deposits by simulated rainfall and the control of eyespot disease by winter wheat. Crop Prot 8:373–379

Cooke BK, Jordan VWL, Hislop EC, Western NM (1993) Mechanism of control of eyespot disease in winter wheat by the fungicide prochloraz. Crop Prot 12:279–283

Coupland D, Taylor WA, Caseley JC (1978) The effect of site of application on the performance of glyphosate on Agropyron repens and barban, benazoylprop-ethyl and difenzoquat on Avena fatua. Weed Res 18:123–128

Davies PJ, Drennan DSH, Fryer JD, Holly K (1967) The basis of the differential phytotoxicity of 4-hydroxy-3,5-di-iodobenzonitrile. I. The influence of spray retention and plant morphology. Weed Res 7:220–233

Dayawon MM, Bohn JA, Striebel SM, Rao SR, Sandbrink JJ, Becher DZ, Petroff LJ, Romenesko DJ, Ekeland RA, Difate VG (1989) Improved herbicide formulations and their use. International patent application no PCT/US89/02570

de Ruiter H, Uffing AJM, Meinen E, Prins A (1990) Influence of surfactants and plant species on leaf retention of spray solutions. Weed Sci 38:567–572

Field RJ, Bishop NG (1988) Promotion of stomatal infiltration of glyphosate by an organosilicone surfactant reduces the critical rainfall period. Pestic Sci 24:55–62

Furmidge CGL (1962) Physicochemical studies on agricultural sprays. IV. The retention of spray liquids on leaf surfaces. J Sci Food Agric 13:127–140

Gaskin RE, Holloway PJ (1992) Some physicochemical factors influencing foliar uptake enhancement of glyphosate-mono (isopropylammonium) by polyoxyethylene surfactants. Pestic Sci 34:195–206

Gauvrit C, Dufour JL (1990) Effect of adjuvants on herbicidal action. I. Effects of a mixture of adjuvants on diclofop-methyl retention and penetration in wheat and ryegrass. Agronomie 11:27–34

Grayson BT, Webb JD, Pack SE, Edwards D (1991) Development and assessment of a mathematical model to predict foliar spray deposition under laboratory spraying conditions. Pesticide Science 33:282–304

Grayson BT, Pack SE, Edwards D, Webb JD (1993) Assessment of a mathematical model to predict spray deposition under laboratory track spraying conditions. II. Examination with further plant species and diluted formulations. Pestic Sci 37:133–140

Hibbitt CJ (1969) Growth and spray retention of wild oat and flax in relation to herbicidal selectivity. Weed Res 9:95–107

Holloway PJ (1993) Adjuvants for agrochemicals: why do we need them? Medelingen van de Faculteit Landbouwwetenschappen Rijksuniversiteit Gent 58/2a:125–140

Holloway PJ, Wong WW-C, Partridge HJ, Seaman D, Perry RB (1992) Effects of some nonionic polyoxyethylene surfactants on uptake of ethirimol and diclobutrazol from suspension formulations applied to wheat leaves. Pestic Sci 34:109–118

Holly K (1976) Selectivity in relation to formulation and application methods. In: Audus LJ (ed) Herbicides: physiology, biochemistry, ecology, vol 2. Academic, London, pp 249-277

MacIsaac S, Paul RN, Devine MD (1991) A scanning electron microscope study of glyphosate deposits in relation to foliar uptake. Pesticide Science 31: 53–64

Manthey FA, Szelezniak EF, Anyszka ZM, Nalewaja JD (1992) Foliar absorption and phytotoxicity of quizalofop with lipid compounds. Weed Sci 40:558–562

McWhorter CG, Barrentine W (1988) Spread of paraffinic oil on leaf surfaces on Johnsongrass (Sorghum halpense). Weed Sci 14:645–658

Murphy GJ, Policello, Ruckle RE (1991) Formulation considerations for trisiloxane based organosilicone adjuvants. Proceedings of the Brighton crop protection conference, Weeds 1991, pp 355–362

Nalewaja JD (1986) Seed oils with herbicides. Mededelingen van de Faculteit Landbouwwetenschappen Rijksuniversiteit Gent 51(2a):301–310

Nalewaja JD, Palczynski J, Manthey FA (1990) Imazethapyr efficacy with adjuvants and environments. Weed Technol 4:765–770

Nalewaja JD, Matysiak R, Freeman TP (1992) Spray deposit residual of glyphosate in various carriers. Weed Sci 40:576–589

Nassetta M, Remedi MV, de Rossi RH (1991) Effect of surfactants on the solubility of herbicides. J Agric Food Chem 39:1175–1178

Riederer M, Schönherr J (1990) Effects of surfactants on water permeability of isolated plant cuticles and the composition of their cuticular waxes. Pestic Sci 29:85–94

Roggenbuck FC, Penner D, Burow RF, Thomas B (1993) Study of the enhancement of herbicide activity and rainfastness by an organosilicone adjuvant utilizing radiolabelled herbicide and adjuvant. Pestic Sci 37:121–125

Scher HB (1984) Innovations and developments in pesticide formulations. In: Cross B, Scher HB (eds) Pesticide formulations innovations and developments. American Chemical Society, Washington DC, pp 1–5 (ACS symposium series 371)

Schönherr J, Riederer M (1988) Desorption of chemicals from plant cuticles: evidence of asymmetry. Arch Environ Contaminat Toxicol 17:13–19

Schott JJ, Dufour JL, Gauvrit C (1991) Effects of adjuvants on herbicidal action. III. Effects of petroleum and rape oils on diclofop-methyl action on ryegrass. Agronomie 11:27–34

Schreiber L, Schönherr J (1992) Analysis of foliar uptake of pesticides in barley leaves: role of epicuticular waxes and compartmentation. Pestic Sci 36:213–221

Silcox D, Holloway PJ (1986) A simple method for the removal and assessment of foliar deposits of agrochemicals using cellulose acetate film stripping. Aspects Appl Biol 11:13–17

Stevens PJG, Gaskin RE, Hong S, Zabkiewicz JA (1991) Contributions of stomatal infiltration and cuticular penetration to enhancement of foliar uptake by surfactants. Pestic Sci 33: 371–382

Stock D (1991) Effects of adjuvants on biological activity of foliar-applied pesticides. Proceedings of the Brighton crop protection conference, Weeds 1991, pp 315–322

Stock D, Edgerton BM, Gaskin RE, Holloway PJ (1992) Surfactant-enhanced foliar uptake of some organic compounds: interaction with two model polyoxyethylene aliphatic alcohols. Pestic Sci 34:233–242

Stock D, Holloway PJ, Grayson BT, Whitehouse P (1993) Development of a predictive uptake model to rationalize selection of polyoxylethylene surfactant adjuvants for foliage-applied agrochemicals. Pesticide Science 37:233–245

Temple RE, Hilton HW (1963) The effect of surfactants on the water solubility of herbicides and the foliar phytotoxicity of surfactants. Weeds 11:297–300

Urvoy C, Gauvrit C (1991) Seed oils as adjuvants: penetration of glycerol trioleate, methanol oleate and diclofop-methyl in maize leaves. Proceedings of the Brighton crop protection conference, Weeds 1991, pp 337–342

Watanabe T, Yamaguchi I (1991) Wettability characteristics of crop leaf surfaces. J Pestic Sci 16: 651–663

Wirth W, Storp S, Jacobsen W (1991) Mechanisms controlling leaf retention of agricultural spray solutions. Pestic Sci 33:411–420

3 Influence of Formulation and Adjuvants on the Foliar Location and Physical Form of the Active Ingredient

R. H. Falk

3.1 Introduction

The efficacy of a foliar-applied herbicide ultimately depends on the amount of active ingredient that reaches a site of action. Herbicide molecules must pass from the outer surface of the plant leaf, through the cuticle, and into the underlying tissue. A number of interrelated factors influence this efficacy and may also lead to herbicide specificity. For example, the physical and chemical nature of the cuticle, the nature of the herbicide formulation and the environmental history of the plant all appear to affect herbicide efficacy. We still have a poor understanding of the influence of formulation variables and adjuvants on the foliar form of herbicide deposits, on their distribution, and on any relationships between physical form, distribution, and biological activity.

This paper, after a brief review of general leaf surface features and the chemical nature of the plant cuticle, surveys the results of studies dealing with the influence of adjuvants on the foliar form of herbicide deposits, how adjuvants affect deposit distribution on the leaf surface, and relationships between form, distribution and, phytotoxicity.

3.2 General

The aerial surfaces of all higher plants have a cuticle. This relatively thin structure (0.1 µm in some species of *Prunus* to more than 13 µm in *Olea lancea*) is composed of an insoluble cutin matrix and soluble waxes (Martin and Juniper 1970; Price 1982; Holloway 1993).

3.3 Waxes

Some waxes are embedded within the cutin matrix and are referred to as cuticular waxes. Baker (1982) reports that cuticular waxes are principally fatty acids; however, their exact composition remains uncertain.

Other waxes are found on the surface of the cutin matrix and are referred to as epicuticular waxes. Epicuticular waxes have been studied by many individuals (e.g., Baker 1982; Gülz et al. 1987; Tulloch and Hoffman 1973; Bianchi 1985) and are found to possess a diverse composition that varies according to species. Holloway (1993) has listed classes of wax constituents on the basis of their frequency of occurrence. These are summarized in Tables 1 and 2.

Wax constituents often give rise to a microstructure which profoundly affects the wettability of foliage, which in turn affects how foliar-applied herbicides distribute themselves on a leaf surface. Various wax compositions appear as granules, rods, plates, crusts, aggregates, or a smooth, amorphous surface (Amelunxen et al. 1967). Epicuticular waxes may not be uniformly distributed on the leaf surface, and there may be differences between the waxes of the abaxial and adaxial surfaces of

Table 1. Most common plant epicuticular wax constituents (Holloway 1993)

Class	Formula	Homologue range
n-Alkanes	$CH_3–(CH_2)_n–CH_3$	Odd $C_{17}–C_{35}$ Often C_{29} or C_{31}
n-Alkyl-monoesters	$CH_3–(CH_2)_n–CO–O–(CH_2)_n–CH_3$	Even $C_{32}–C_{72}$
n-Aldehydes	$CH_3(CH_2)_n–CHO$	Even $C_{16}–C_{34}$ often C_{26} or C_{28}
n-1-Alkanols	$CH_3–(CH_2)_n–OH$	Even $C_{18}–C_{36}$ often C_{26} or C_{28}
n-Alkanoic acids	$CH_3–(CH_2)_n–COOH$	Even $C_{14}–C_{36}$ often C_{26} or C_{28}

Table 2. Less common plant epicuticular wax constituents (Holloway 1993)

Class	Formula	Homologue range
n-Ketones	$CH_3–(CH_2)_n–\underset{*}{\overset{\overset{O}{\parallel}}{C}}–(CH_2)_n–CH_3$	Odd $C_{23}–C_{38}$ Often C_{29} or C_{31}
n-sec-Alcohols	$CH_3–(CH_2)_n–\underset{*}{\overset{\overset{OH}{\mid}}{CH}}–(CH_2)_n–CH_3$	Odd $C_{21}–C_{33}$ Often C_{29} or C_{31}
β-Diketones	$CH_3–(CH2)_n–\overset{\overset{O}{\parallel}}{C}–\underset{*}{CH_2}–\overset{\overset{O}{\parallel}}{C}–(CH_2)_n–CH_3$	Odd $C_{29}–C_{33}$
Pentacyclic triterpenoids	Triterpenoids, e.g., β-amyrin Triterpenoid acids, e.g., ursolic acid	

*Different positional isomers occur

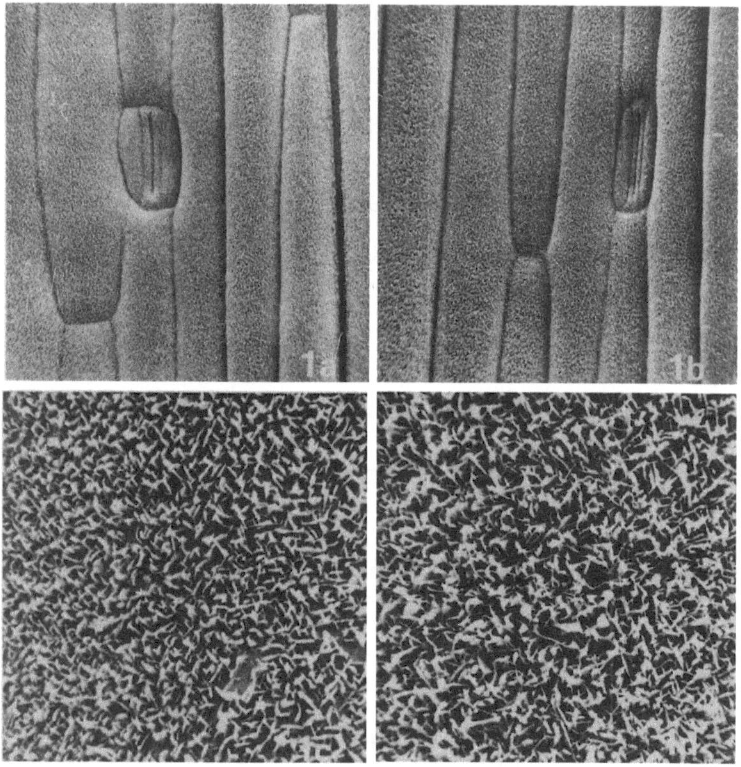

Fig. 1a–d. The leaf surfaces of *A. fatua* L. **a** Upper surface, × 350. **b** Lower surface, × 350. **c** Upper surface, × 500. **d** Lower surface, × 3500. (From Harr et al. 1991)

a particular leaf. The leaf surface of *Avena fatua* L. (wild oat) with its characteristic platelike epicuticular wax is shown in Fig. 1. *Brachiaria plantaginea* (Link) Hitchc. (plantain signalgrass), shown in Fig. 2, has a rather smooth lower leaf surface. Its upper surface, however, is well decorated with wax structures. Wax ornamentation is often less conspicuous over guard and accessory cells and may allow these surfaces to be more easily wetted. The leaf surfaces of *Abutilon theophrasti* Medik. (velvet leaf), a dicot with amorphous epicuticular wax is shown in Fig. 3.

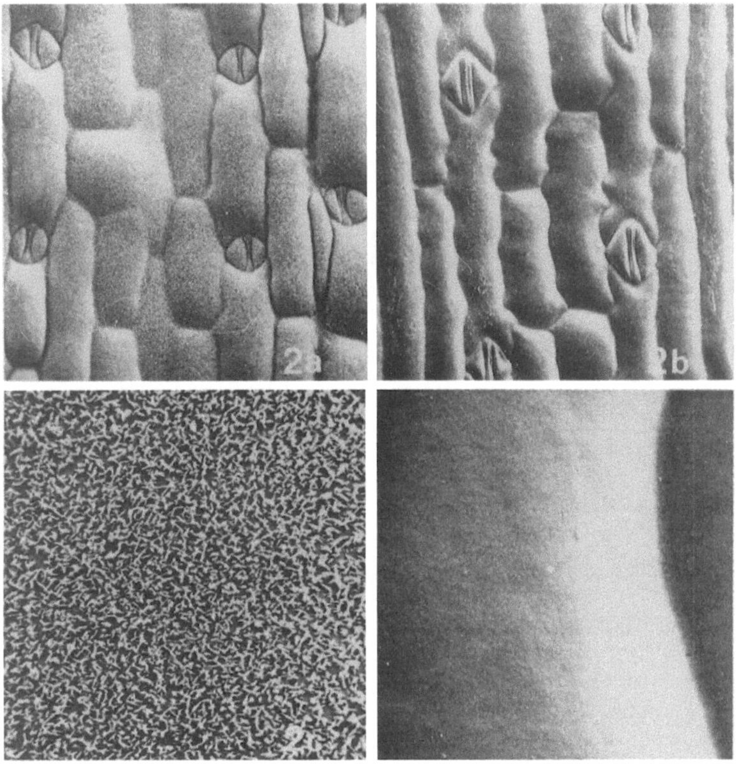

Fig. 2a–d. The leaf surfaces of *B. plantaginea* (Link) Hitchc. **a** Upper surface, × 350. **b** Lower surface, × 350. **c** Upper surface, × 3500. **d** Lower surface, × 3500. (From Harr et al. 1991)

Gas chromatographic analyses of the total chloroform extractable epicuticular waxes of these three species are shown in Fig. 4 and reveal the substantial differences in composition that they present. Wax microstructure can often be related to chemical composition (Baker 1982).

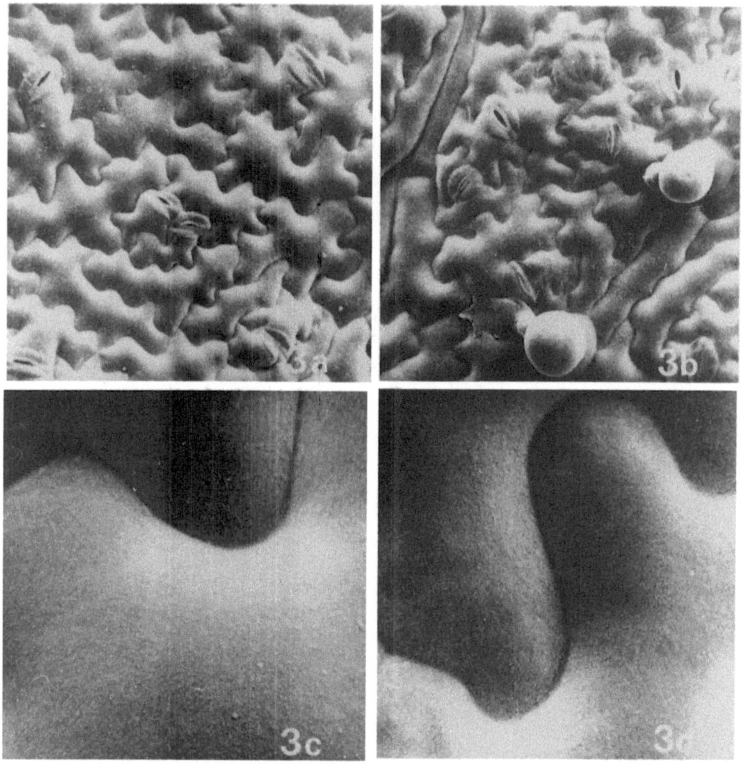

Fig. 3a–d. The leaf surfaces of *A. theophrasti* Medik. **a** Upper surface, × 350. **b** Lower surface, × 350. **c** Upper surface, × 3500. **d** Lower surface, × 3500. (From Harr et al. 1991)

3.4 Cutin

Cutin is a polyester based on a series of hydroxylated fatty acids of most commonly 16 or 18 carbon chain lengths (Kolattukudy 1980, 1981; Holloway 1982, 1993). The relative proportion of C_{16}–C_{18} fatty acids varies according to species; however, there appears to be no obvious chemotaxonomic relationship between cutin composition and species (Chamel 1986).

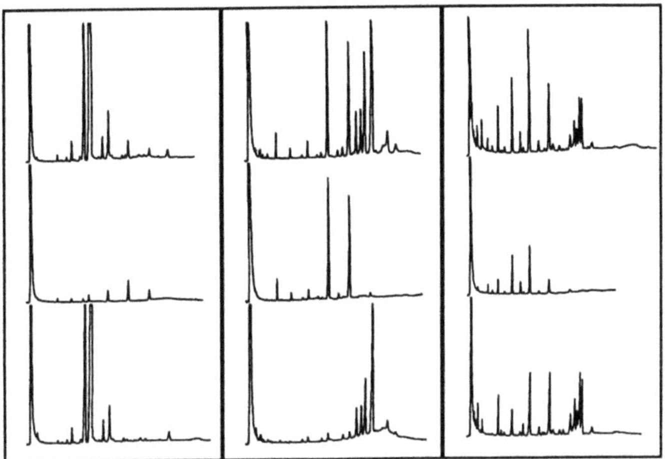

Fig. 4. Gas chromatograms of epicuticular waxes. *Above*, total wax extract; *middle*, hexane soluble fraction; *below*, total-hexane soluble fraction

The cuticle is not a homogeneous layer. Its outer surface is quite lipophilic, and it becomes increasingly hydrophilic as one moves toward the underlying epidermal cells. The presence of polar pathways through the cuticle has been indirectly suggested (Schönherr and Bukovac 1970) and that some preferential movement of herbicides may occur in these channels. Relatively large channels can be seen in some cuticles by light microscopy. Miller (1985) has reported the occurrence of such channels in the leaves of 26 taxa among 13 families. In a second study Miller (1986) added 37 taxa among 19 other families to those with such channels. Their significance in herbicide uptake has not been established. Perhaps Crafts (1961) describes it best when he suggests that the cuticle is spongelike with discontinuous polar and nonpolar regions.

Carbohydrate polymers associated with the cell wall are often seen to extend into the cuticle (Hallam and Juniper 1971) and may represent transcuticular pathways.

Cuticles can also be described as polyelectrolytes. When isolated tomato fruit cuticles were titrated potentiometrically using various bases, Schönherr and Bukovac (1973) observed three dissociable groups in the pH ranges 3–6, 6–9, and 9–12. The first group was tentatively assigned to -COOH groups of pectic materials and protein embedded in the cuticle, the second to nonesterified -COOH groups of the cutin polymer, and the third to phenolic -OH groups and to a small amount of $-NH_3^+$ groups of proteins. Schönherr and coworkers (Schönherr and Bukovac 1973; Schönherr and Huber 1977) view the cuticle as a three-dimensional ion exchange matrix where, at physiological pH, the cuticle is negatively charged, and the majority of the exchange sites are occupied by Ca^{+2}.

Partitioning into the cuticle is strongly influenced by the chemical nature of applied herbicides. The uptake of weak acids such as 2,4-D, 2,4,5-T, and 1-naphthaleneacetic acid increases at lower pH values, indicating that nondissociated species are better able to enter (Riederer and Schönherr 1984) since ionized forms of weakly acidic herbicides are subject to Donnan exclusion.

3.5 Trichomes

The role of trichomes in foliar penetration has received rather little attention. Hull et al. (1975) have pointed out that trichomes may represent a distinct portal of entry. Benzing and Burt (1970) studied the uptake of ^{45}Ca and ^{65}Zn in 20 species of Bromeliaceae. They found the greatest uptake in species covered with trichomes. In a large study of 35 weed species Ormrod and Renney (1968) reported that only 9 have pubescences sufficient to influence retention or penetration of sprays; these are: *Lappula echinta* Gilib. (blue-bur), *Cerastinum vulgatum* L. (common mouse-ear chick-weed), *Lychnis alba* Mill. (White campion), *Hypochaeris radiata* L. (spotted catsear), *Centaurea diffusa* Lam. (diffuse knapweed), *Raphanus raphanistrum* L. (jointed charlock), *Capsella bursa-pastoris* (L.) Medic. (shepherd's purse), *Galeopsis tetrahit* (Boenn.) Lej. & Court. (common hemp-nettle), and *Ranunculus repens* L. (Creeping buttercup). To this list one should add *Eremocarpus setigerus* (Hook.) Benth. (turkey mullein). Hess et al. (1974) found that large drops from foliar sprays shatter on impact with its trichomes, greatly affecting the amount of [(4-chloro-*o*-tolyl)oxy]acetic acid (MCPA) that reached the true epidermis.

3.6 Stomata

Leaf stomata are apertures formed by two guard cells (modified epidermal cells). Changes in the turgor of these guard cells regulates the area of the stomatal aperture. When open, stomata lead to substomatal chambers surrounded by mesophyll tissue and thus effect a portal through which gas exchange between leaf cells and the external environment may occur.

Stomata have been frequently implicated in the foliar uptake process. While their involvement in foliar absorption is well documented, the degree and mechanism remain unclear. Some early studies revealed that foliar penetration is greater through stomatous than astomatous surfaces (Sargent 1965), and that penetration is correlated with stomatal frequency (Jyung et al. 1965). This circumstantially led to the idea that stomata provide direct pathways for penetration. Schönherr and Bukovac (1972) and Greene and Bukovac (1974) carefully examined the penetration of stomata by liquids as a function of surface tension, wettability, and stomatal morphology. Liquids having a surface tension less than 30 dyn/cm gave zero contact angle and infiltrated stomata spontaneously while liquids having a surface tension greater than 30 dyn/cm did not wet the leaf surface and failed to infiltrate leaf stomata. It is clear that the surface tensions provided by surfactants used in most agrochemical sprays do not provide significant stomatal infiltration.

Relatively few surfactants will reduce the surface tension of aqueous solutions below 30 dyn/cm. Zabkiewicz et al. (1988), Field and Bishop (1988), and Stevens et al. (1992) have pointed out that selected organosilicone surfactants are able to effect such surface tension reductions and allow stomatal flooding. Stevens et al. (1992) remind us also that stomatal infiltration can occur only during the brief period after application while spray deposits are still liquid. After this, cuticular penetration remains the sole uptake pathway.

Interestingly, the degree of stomatal opening has been found to be of little importance. Spontaneous infiltration of stomata takes place when the contact angle of an applied material is smaller than the wall angle of the aperture wall. Cuticular ledges present at the entrance to the outer vestibule and between the inner vestibule and substomatal chamber result in very small if not zero wall angles and thus play a major role in excluding water from the intercellular space of leaves.

3.7 Formulation Effects on Herbicide Absorption

Herbicides are almost never applied alone or as simple solutions. Instead, manufacturers add adjuvants, a heterogenous group of substances, which in some manner enhance the desired action or performance of the herbicide. These types of adjuvants are formulation adjuvants, and the properties that they affect are: solubility, volatility, specific gravity, corrosiveness, shelf life, compatibility with other products, and characteristics such as spreading and penetration.

The end user of herbicides often chooses to add additional adjuvants, and while a number are available for specific purposes, the most common are the "activation adjuvants", i.e., surfactants, oils, and oil-emulsifier concentrates.

Adjuvants may influence the activity of a herbicide at one or more sites: within the formulation itself, on the cuticle surface, within the cuticular layers; and at the plasmalemma; within the cytoplasm. The following addresses the first and second of these possibilities: the formulation itself and the cuticle surface.

3.7.1 Do Surfactants Solubilize Epicuticular Waxes?

Takeno and Foy (1974) reported severe erosion of the epicuticular waxes of *Gossypium hirsutum* L. (cotton) after 72 h of exposure to polysorbate surfactant. *Sida spinosa* L. (prickly sida) was much less affected by surfactant exposure.

Rawlinson et al. (1978) working with four cultivars (Primor, Eurora, Victor and Rapol) of *Brassica napus* L. ssp *oleifera* (rape) reported that commercial formulations of the herbicides dalapon (sodium 2,2-dichloropropionate), carbetamide [*N*-ethyl-2(phenylcarbamoyloxy)propionamide] and propyzamide [3,5-dichloro-*N*-(1,1-dimethylpropynl)benzamide] alter the topography of leaf epicuticular wax. In particular, dalapon and carbetamide tended to decrease the area of leaf surface occupied by upright rodlets of wax and flatten the underlying dendritic wax platelets of cultivars Primor and Eurora, both low erucic acid cultivars. On the high erucic acid cultivars, Victor and Rapol, the effects were less noticeable after dalapon treatment and did not occur after carbetamide treatment.

Kuzych and Meggitt (1983) reported that when the leaves of *B. oleracea* L. var. Copenhagen market (cabbage), *Datura stramonium* L. (jimsonweed) and *A. theophrasti* Medik. (velvetleaf) were dipped halfway into the surfactants, petroleum oil distillates, X-77 (alkylarylpolyoxyethylene-glycol-free fatty acid-isopropanol mixture), and d'limonene [1-methyl-4-(1-methylethenyl)cyclohexene] to provide a margin where treated and untreated regions could be compared, each surfactant caused a pattern of disruption. In general, the greatest surface disruption was caused on cabbage by X-77, followed by the petroleum distillates and d'limonene.

Partial dissolution of epicuticular wax deposits on *A. fatua* L. (wild oat) was reported by Whitehouse et al. (1982) following treatment with formulated ethofumosate [(±)-2-ethoxy-2,3-dihydro-3,3-dimethyl-5-benzofuranyl methanesulfonate], S-ethyl dipropylthiocarbamate, and dalapon (sodium 2,2-dichloropropionate).

Noga et al. (1991) examined the effects of Triton X-100 (octylphenoxy polyethoxy ethanol) on the epicuticular waxes of *B. oleracea* L. var. *gemmifera* DC, *B. oleracea* var. *gongylodes* L. and *Triticum aestivum* L. They found that Triton X-100 causes pronounced modifications of *B. oleracea* var. *gongylodes* waxes and slight changes in the waxes of *Triticum aestivum*. The same investigators reported that while a short dip in water caused no wax structural changes, the application of a 1-μl droplet of water to a leaf surface for 2 min caused significant changes.

Wilkinson and Mayeux (1987) have proposed a mechanism for epicuticular wax loss by rainfall from *Isocoma coronopifolia* (Grey) Greene (common goldenweed) and *Isocoma drummondii* (T & G) Greene (Drummond's goldenweed). The epicuticular waxes of these species consist primarily of free fatty acids and alcohols of low molecular weight. They suggest that the free fatty acids form salts with cations available in alkaline dust on the leaf surface. The resulting short-chain soaps would be sufficiently water soluble to be removed by rain.

In a study on effects of oxyethylene chain length of octylphenoxy (OP) surfactants, Knoche and Bukovac (1991) examined the effect of a series of octylphenol surfactants with oxyethylene (EO) chain lengths of 5, 7–8, 9–10, 16, and 30 (Triton X surfactants) on the uptake of gibberellic acid by leaves of *Prunus cerasus* cv Montmorency. They found that penetration through the abaxial leaf surface was linearly related to the logarithm of the oxyethylene content of the surfactant molecule.

Enhancement of penetration was greatest with surfactants having a hydrophile chain length of 9–10 EO units or less. Triton X-165 (OPE 16) was next most efficacious. Triton X-305 (OPE 30) did not effect gibberellic acid penetration at all when compared to a control.

Knoche et al. (1992) in a study involving surfactant-induced phytotoxicity looked at a larger number of species: *B. oleracea* var. *gemmifera* DC cv Lunst (brussels sprout), *B. oleracea* var. *gongylodes* L. cv Express Forcer (kohlrabi), *Vicia faba* L. cv Hang Down (broad bean), *Zea mays* L. cv Sigma (maize), *B. napus* var. *napus* L. (rape), *Hordeum vulgare* L. cv Tapir (barley), and *Triticum aestivum* L. cv Okapi (wheat). They treated these species with commercial preparations of a series of OP surfactants with EO chain lengths of 5, 7.5, 9.5, 16, and 30 (Triton X surfactants) and a series of nonylphenol surfactants with EO chain lengths of 7, 10, 14, 20, and 30 (Lutensol AP surfactants). They found that the epicuticular wax fine structure of the leaf surfaces of *B.* species is a critical component contributing to its unique toxicity pattern after exposure to surfactants.

Bukovac et al. (1983) examined the leaf surfaces of *B. oleracea* var. *capitata* L. (cabbage), *B. oleracea* L. var. *italica* (broccoli) and *Pyrus communis* L. cv Bradford (pear) with a scanning electron microscope. They reported that the surface fine structure of these species is not significantly deformed by Tween 20 (polyoxyethylene sorbitan monolaurate), Regulaid, or Triton B-1956 (modified phthalic glycerol alkyl resin). Plants treated with X-77 (alkylarylpolyoxyethylene-glycol-free fatty acid-isopropanol mixture), d'limonene (1-methyl-4-(1-methylethenyl)cyclohexene), and Triton CS-7 (blend of alkylaryl polyethoxylate and sodium salt of alkyl sulfonated alkylate) exhibited a subtle deformation of wax fine structure. Triton X-100 (alkylphenoxy polyethoxy ethanol), Vatsol OT (sodium dioctylsulfosuccinate) and Hyamine 2389 (methyl dodecyl benzyl trimethyl ammonium chloride, methyl dodecyl xylylene bis trimethyl ammonium chloride) treatment of these species caused obvious wax fine structure deformation.

Wolter et al. (1988) working on a wax regeneration study involving *B. oleracea* var. *gongylodes* L. and *Beta vulgaris* var. *altissima* Doell, reported that Triton X-100 (octylphenoxy polyethoxy ethanol) applied as 1μ-l droplets caused significant alterations of epicuticular wax in *Brassica*.

Noga et al. (1987) examining changes in leaf micromorphology by surfactants reported that in *B. oleracea* L. var. *gongylodes* and *Beta vulgaris* var. *altissma* Doell, application of the surfactants Triton X-100 (octylphenoxy polyethoxy ethanol) and Triton X-114 (octylphenoxy polyethoxy ethanol) caused a disruption of epicuticular wax on *B*. *Beta vulgaris* does not have crystalline wax and the authors reported no solubilization of epicuticular wax for this species.

Stevens and Bukovac (1987a,b) in a study on OP surfactants mention that scanning electron microscopic observation of *Zea mays* L. (maize) leaf surfaces exposed to a series of octylphenol surfactants with EO chain lengths of 5, 7.5, 9.5, 12.5, 16, 30, and 40 (Triton X-45, X-114, X-100, X-102, X-164, X-305, and X-405 surfactants) did not reveal any visible modifications of epicuticular wax structure. They also calculated that none of the surfactants examined in their study would solubilize more than 0.1% of the epicuticular wax underlying a droplet deposit in maize. This calculation was based on the solubility of pentacosane in aqueous surfactant solutions and on the assumption that the majority of the epicuticular wax components of corn are apolar.

Stevens and Baker (1987), in a large study involving *Malus pumila* cv Yarlington Mill (apple), *Gossypium hirsutum* cv Delta Pine (cotton), *Vicia faba* L. cv Maris Bead (field bean), *Zea mays* L. cv LG11 (maize), *Pisum sativum* L. cv Meteor (pea), *B. napus* L. cv Rafal (rape), *Spinacia oleracea* L. cv Supergrade (spinach), *Beta vulgaris* L. cv Nomo (sugar beet), *Fragaria ananassa* cv Red Gauntlet (strawberry), *Citrus limon* Burm. (lemon), and *Solanum tuberosum* f. spp. *tuberosum* cv King Edward (potato) and the surfactant, Ethylan TU (nonylphenol 8 ethoxylate) found that wax morphology differs widely among the above species, and that the surfactant effects no visible modification of the epicuticular wax.

De Ruiter et al. (1988) in a study examining Renex 688 (polyoxyethylene nonylphenol), Ethomeen T25 (polyoxyethylene 15 tallow amine), and technical glyphosate reported that the crystalline epicuticular wax of *Triticum aestivum* L. cv Arminda (winter wheat) seemed unimpaired 24 h after application of surfactant.

Falk et al. (1993) examined the leaf surfaces of *Ipomoea purpurea* L. Roth. (tall morningglory), *Spirodela polyrrhiza* L. (giant duckweed), and *Portulaca oleracea* L. (common purslane) after treatment with Herbex (polymeric polyhydroxy acids), Andogen 462 (dimethyl dicoco ammonium chloride), Bardac 22 (dioctyl dimethyl ammonium chloride), Emcol CC-47 (quaternary ammonium compound), Ethomeen T25 (polyoxyethanol 15 tallow amine), Antarox Co 730 (nonylphenol ethoxylate), Ethylan A3 (polyethylene glycol ester of oleic acid), Triton X-100 (alkylphenoxy polyethoxy ethanol), Tronic (a mixture of alkylarylpolyoxyethylene glycols, alkyl amino acetate, alkylarylsulfonates, and polyhydric alcohol) and found no morphological alteration of epicuticular waxes.

Reports of alteration of wax morphology by surfactants, particularly Triton X-100, are commonly associated with experiments involving *Brassica* species. When gas chromatograms (Fig. 5) of the epicuticular waxes of *B. napus* are compared to those of *I. purpurea*, *S. polyrrhiza*, and *P. oleracea*, one notes that *Brassica* has rather late eluting components relative to the other three species. Recalling that Knoche et al. (1992) proposed that the epicuticular wax fine structure of leaf surfaces of *Brassica* species is a critical component contributing to the unique toxicity pattern observed after its application, perhaps the chemistry of

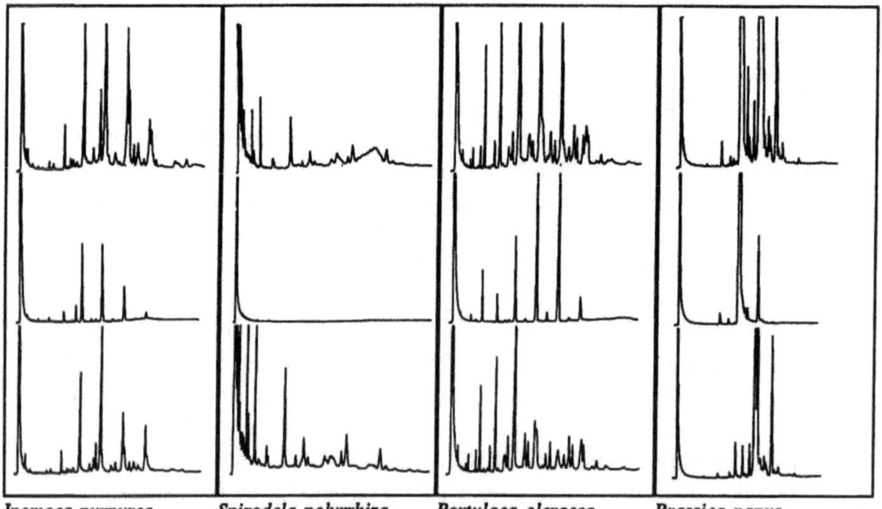

Fig. 5. Gas chromatograms of epicuticular waxes. *Above*, total wax extract; *middle*, hexane soluble fraction; *below*, total-hexane soluble fraction

the epicuticular waxes of *Brassica* allows alteration of its morphology by surfactants more easily.

3.7.2 Crystalline or Amorphous Residues

Takeno and Foy (1974) in a study of the effects of polysorbate surfactants on the ultrastructure of the leaf surfaces of *Gossypium hirsutum* L. (cotton) and *Sida spinosa* L. (prickly sida) noted that crystalline deposits were observed on the leaves when technical herbicide-ethanol solution was applied; amorphous deposits were present when a wettable powder suspension was applied.

Hess et al. (1987) showed that formulation components of emulsifiable concentrates can influence the physical form of herbicides after spraying onto plant surfaces. The formulation of flamprop-ethyl (ethyl *N*-benzoyl-*N*-(3-chloro-4-fluorophenyl)-2-aminopropionate) that yielded crystalline deposits on the leaf surface of *A. fatua* L. (wild oat) was not absorbed as efficiently as a second formulation which did not crystallize on the leaf surface. This second formulation contained less a. i. (8% versus 13%) and more emulsifier (12% versus 6.5%) than the first.

De Ruiter et al. (1988) mentioned in their study on the mode of action of a nonionic and cationic surfactant that hydrated Renex 688 [polyoxyethylene (8) nonylphenol] exists as a gel whereas hydrated Ethomeen T25 (polyoxyethylene 15 tallow amine) exists as a soft solid form, and that the differences in uptake and physical form of these two surfactants support the conclusion of Anderson and Girling (1983) that the uptake of surfactants is at least partly related to their physical form.

In a very large study by Baker et al. (1992) the uptake and translocation of 26 chemicals (herbicides, fungicides, growth regulators, insecticides, and model compounds) formulated with and without the surfactant Ethylan TU (nonylphenol ethoxylate) were determined after spraying onto leaves of *Zea mays* L. cv LG11 (maize), *B. napus* L. cv. Rafal (rape), *Beta vulgaris* L. cv Nomo (sugar beet) and *Fragaria ananassa* cv Red Gauntlet (strawberry). Chemicals penetrated more readily into waxy leaves of rape and strawberry than into the less waxy sugar beet. The lowest uptake rates were observed for maize. The authors classified the effects of Ethylan TU on uptake rates into four groups. They pointed out that the lack of a clear, simple relationship between the physicochemical properties of the 26 chemicals and their uptake and redistribution patterns confirms the complexity of the uptake process. Some chemicals formed crystals, and it is possible that some of the variation in uptake arises from differences in the physical form of the dried deposit. Crystals make poor contact with the cuticle, which undoubtedly contributes to their low uptake rates. In their study chemicals having a melting point of greater than 200°C formed crystalline deposits regardless of whether formulated with or without surfactant. Chemicals with a melting point between 150° and 200°C formed crystalline deposits in the absence of surfactant but amorphous deposits in the presence of surfactant. Finally, chemicals with a melting point between 40° and 135°C formed amorphous deposits in both the presence and the absence of surfactant.

The greatest uptake rates were found with chemicals having low melting points and lipophilicity. These formed amorphous deposits which partially merged with epicuticular wax. That is, the dried deposits are likely to consist of a water surfactant gel which retains much of the a. i. in relatively high concentration in a partially hydrated form and maintains an intimate contact between the deposit and the leaf cuticle. Diffusion rates of polar compounds would be expected to be low; highly lipophilic compounds would be expected to bind tightly to the cutin matrix. Finally, the authors point out that alternative mechanisms are required to account for the effective performance of water soluble chemicals such as glyphosate.

An earlier and less expansive study was carried out by Baker and Hunt (1988) using the same plant species and surfactant. It was reported here that, almost without exception, spray droplets containing surfactant dried down to form an annulus at the periphery of the dried deposit whereas uniform deposits were obtained in the absence of surfactant, particularly for chemicals of medial lipophilicity. The physical form of the deposits when viewed with the scanning electron microscope varied with the polarity of the a. i. Water soluble compounds crystallized as large needles both with and without surfactant. Compounds of median lipophilicity and water solubility such as 1-naphthaleneacetic acid were difficult to see and were presumably absorbed. Compounds of median lipophilicity and low water solubility such as atrazine [2-chloro-4-(ethylamino)-6-(isopropylamino)-*s*-triazine] and simazine [2-chloro-4,6-

bis(ethylamino)-*s*-triazine] formed large amorphous deposits with arrangements of fine crystals projecting out of the deposit. Highly lipophilic materials such as diclofop-methyl {methyl 2-[4-(2, 4-dichlorophenoxy) phenoxy] propanoate} dried as amorphous deposits with and without surfactant.

Baker and Hunt (1988) go on to say that their results support the suggestion of Stevens (1984) that uptake is optimal for compounds of median lipophilicity and remind us that a considerable number of pesticides are of median polarity.

3.7.3 Distribution Patterns on Leaf Surfaces

One potentially important factor that influences the foliar uptake of a chemical is the distribution pattern of that chemical on the plant surface (Falk et al. 1989; Silcox and Holloway 1989).

Hess et al. (1974) examined the distribution pattern of MCPA on *B. oleracea* L. cv *Capitata* (cabbage), *Beta vulgaris* L. (sugar beet), and *Eremocarpus setigerus* Benth. (turkey mullein). They reported that the majority of herbicide accumulated in the depressions above the anticlinal cell walls of the epidermis.

Hess et al. (1981) examined the distribution of the sodium and amine salts of MCPA on *Beta vulgaris* L. (sugar beet) leaves and reported that they appeared as discrete deposits over the anticlinal cell walls. When the sodium salt was applied to *Cynodon dactylon* (L.) Pers. (bermudagrass), the distribution pattern was the same; however, each deposit was significantly smaller. The iso-octyl ester of MCPA coalesced into numerous, small, thick deposits on the leaf surface and was crystalline. They also looked at the distribution of atrazine [2-chloro-4-(ethylamino)-6-(isopropylamino)-*s*-triazine] formulated as a wettable powder and as a flowable. The wettable powder distributed itself much more uniformly than the MCPA salts; however, a uniform layer over the entire leaf surface was not observed. The flowable formulation accumulated mainly at the edges of the original spray droplets. Some crystal formation was noted in this latter formulation. Finally, three different emulsifiable concentrate formulations of propanil [*N*-(3,4-dichlorophenyl)propionamide] were examined. Examination of the first formulation revealed many small, smooth, noncrystalline deposits. The second formulation formed crystalline and noncrystalline structures. The third formulation formed discontinuous, exclusively crystalline structures. The first formulation was found to be most effective in controlling *Echinochloa crus-galli* (L.) Beauv. (barnyardgrass).

Stevens and Bukovac (1987) have examined the nature of deposits of atrazine [2-chloro-4-(ethylamino)-6-(isopropylamino)-*s*-triazine] formulated with a series of *OP* surfactants on the leaf surface of *Zea mays* L. (maize). Marked differences were observed among the surfactant treatments. Atrazine alone crystallized onto the leaf in a small region at the center of the deposit area. This form was modified in the presence of surfactant with an ethylene oxide content of 40 to an amorphous deposit with hairlike structures (presumably crystals) protruding from it. Surfactant with an EO of 16 produced a similar deposit. Surfactant with an EO of 9.5 still showed hairlike structures (crystals), but an underlying amorphous deposit was nearly absent. A diffuse deposit (primarily crystalline) was produced when surfactant with EO of 5 was used.

Steurbaut et al. (1989) in a study dealing with fungicides point out that surfactants can modify the physical form of deposits on the leaf surface, and that physical form may be important for the persistence of the deposit. Liquid and jellylike deposits stick better to the leaf surface than solid or crystalline deposits.

Falk et al. (1989) in a study of the influence of several alkyphenoxy surfactants on the distribution of monosodium methanearsonate on the leaves of *B. oleracea* L. cv *Capitata* (cabbage) reported that the percentage of leaf surface covered bore little correlation to phytotoxicity. No crystalline residue was observed.

MacIsaac et al. (1991) have recently examined the effects of glyphosate acid [*N*-(phosphonomethyl)glycine] and its isopropylamine salt in the presence of Tween 20 (polyoxyethylene sorbitan monolaurate) relative to foliar uptake and deposit morphology on the leaves of *Fagopyrum tataricum* (L.) Gaertn. (tartary buckwheat). When glyphosate alone was applied, it dried to form a combination of large crystals and smooth, amorphous areas. After 72 h, the deposits were almost entirely crystalline. The addition of Tween 20 (polyoxyethylene sorbitan monolaurate) resulted in the formation of a smooth, amorphous deposit upon drying. After 72 h large crystals were detected at the periphery of the deposit.

3.7.4 What About Annulus Formation?

Many investigators (e.g., Hartley and Graham-Bryce 1980; Stevens and Baker 1987; Hart and Young 1987) have noted that an annular ring sometimes forms as individual spray droplets dry down on a leaf surface. Hartley and Graham-Bryce attribute these ringlike deposits to temperature and concentration gradients that arise during solvent evaporation.

Baker et al. (1992) noted that more deposits formed an annulus with than without Ethylan TU (nonylphenol ethoxylate). They used a log linear model to investigate whether there is any relationship between the presence of the surfactant and the formation of an annulus, and whether plant species affects the relationship. They found that differences between plant species is not significant; however, the presence of surfactant is significant. Up to 86% of the 26 chemicals which they surveyed produced an annulus when used with surfactant, compared to 16%–36% without surfactant. Baker et al. calculate that although a high proportion of the formulations which they used formed an annulus at the deposit periphery, contact areas were reduced by only 20%–30%, and that the increased concentration of chemical and surfactant in the annulus clearly compensated for the reduced surface coverage.

3.8 Partition Patterns on a Plant Leaf: A Case Study

One can still ask: (a) how do surfactant/herbicide mixtures partition after application, (b) do nonionic, cationic and anionic surfactants behave differently, and (c) do different herbicides behave differently with a given surfactant? Falk and Schulke (1993) have examined the leaves of *Vicia faba* L. (broad bean) after treatment with the isopropylamine salt of glyphosate [*N*-(phosphonomethyl)glycine] and the free acid of dicamba (3,6-dichloro-2-methoxybenzoic acid) in aqueous solutions to which one of the following surfactants was added: Ethomeen T25 (polyoxyethanol 15 tallow amine–cationic), VSE-100 (dodecylbenzene sulfonic acid, triethanolamine salt–anionic), Silwet L-77 (polyalkylene oxide modified polydimethyl siloxane–nonionic silicone), and Triton X-100 (alkylphenoxy polyethoxy ethanol–conventional nonionic). Falk and Schulke used cryoscanning electron microscopy in connection with

X-ray microanalysis, which has proven to be an excellent technique for the study of partitioning of mixtures of herbicides and surfactants (Hess et al. 1974, 1975, 1981; Falk and Schulke 1993). Secondary electron mode images reveal information about the surface structure of a sample; electron bombardment also yields characteristic X-rays which may be used to map the position of particular elements on a surface.

When applied without a surfactant, dicamba and glyphosate have very similar distribution patterns. The secondary electron mode images (Fig. 6) as well as the X-ray mode images, which detect dicamba as chlorine (Fig. 7a) and glyphosate as phosphorous (Fig. 7b) , reveal a rather uniform accumulation of herbicide over the anticlinal cell walls of the leaf epidermis. As the solvent, water, evaporates, remaining solvent and herbicide naturally fill the crevices associated with the anticlinal cell walls.

As formulations dry, the surfactant concentration increases dramatically, and the potential for electrostatic associations of herbicide and surfactant also increases. Such interactions with cationic surfactants and proteins have been reported by Paar and Norman (1965). That such interactions can also occur with glyphosate has been suggested by Wyrill and Burnside (1977). With dicamba and VSE-100 (anionic, Fig. 8), the herbicide residue, while largely over the anticlinal walls, is also seen to occupy some sites on the periclinal walls. As this formulation dries down, at least three negatively charged species are found: an association of surfactant-herbicide, surfactant, and dicamba itself. These might then be repelled somewhat by the net negative charges of the cell walls, plasmalemma, and cytoplasm (Richard and Slife 1979) and end up filling the anticlinal cell wall regions.

With dicamba and Ethomeen, while one can still see residue in the secondary electron mode image (Fig. 9a), clearly less herbicide residue is present (detected as chlorine, Fig. 9b) than when VSE-100 is used (Fig. 8b). An association between the negatively charged dicamba and the positively charged surfactant is possible, producing a nonpolar complex which would be unperturbed by organismal charge and able to meld with cuticular materials. Cationic surfactants are often reported to be effective enhancers of herbicidal activity (Silcox and Holloway 1989; Richard and Slife 1979).

Herbicide residue (detected as phosphorous, Fig. 10b) after application of the glyphosate and VSE-100 formulation is found largely in the

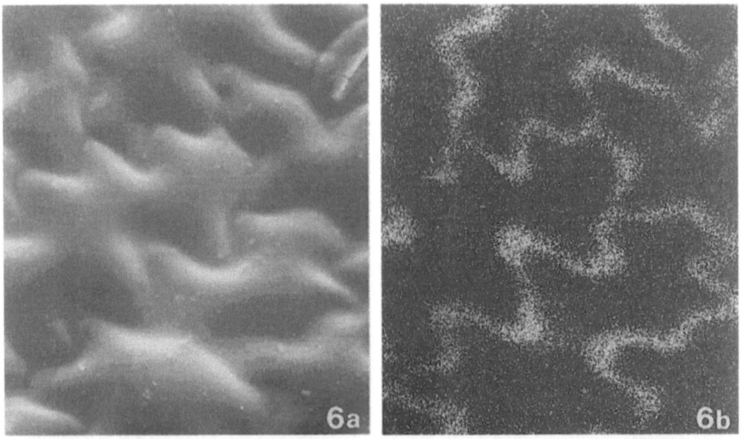

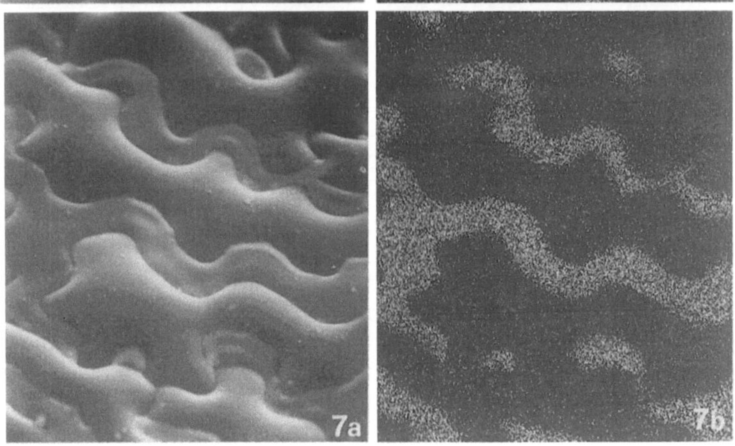

Fig. 6. a Secondary electron image of *V. faba* leaf treated with dicamba alone, × 700. **b** Chlorine X-ray image of **a**. (From Falk and Schulke 1993)

Fig. 7. a Secondary electron image *V.faba* leaf treated with glyphosate alone, × 700. **b** Phosphorous X-ray image of **a**. (From Falk and Schulke 1993)

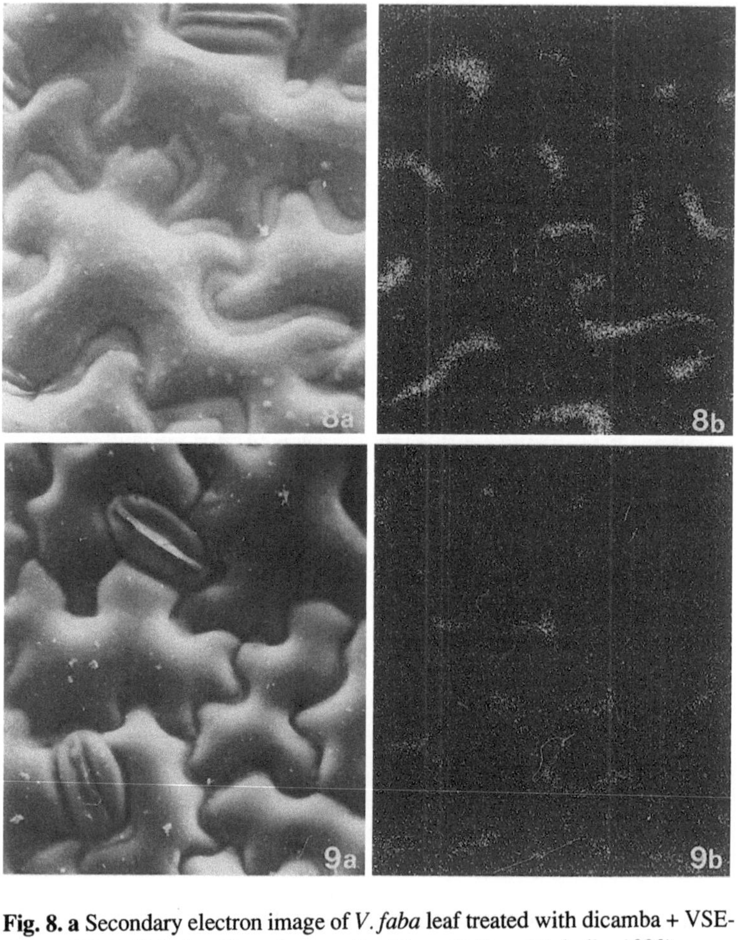

Fig. 8. a Secondary electron image of *V. faba* leaf treated with dicamba + VSE-100, × 700. **b** Chlorine X-ray image of **a**. (From Falk and Schulke 1993)

Fig. 9. a Secondary electron image of *V. faba* leaf treated with dicamba + Ethomeen, × 700. **b** Chlorine X-ray image of **a**. (From Falk and Schulke 1993)

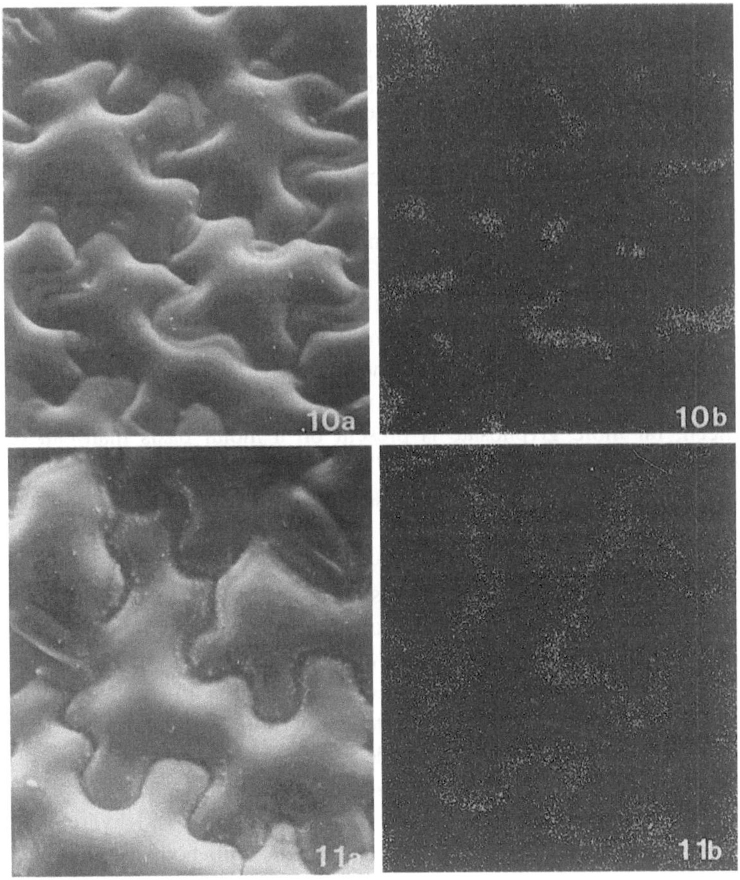

Fig. 10. **a** Secondary electron image of *V. faba* leaf treated with glyphosate + VSE-100, × 700. **b** Phosphorous X-ray image of **a**. (From Falk and Schulke 1993)

Fig. 11. **a** Secondary electron image of *V. faba* leaf treated with glyphosate + Ethomeen, × 700. **b** Phosphorous X-ray image of **a**. (From Falk and Schulke 1993)

anticlinal wall regions of the epidermis (Fig. 10a). Glyphosate and VSE-100 should exhibit little electrostatic interaction beyond the partial positive charge of amino group of glyphosate interacting a bit with the negatively charged surfactant. Overall, the formulation is negatively charged material and will retract to fill the spaces over the anticlinal cell wall regions, perhaps being somewhat repelled by the net negative charge of organismal materials.

In the case of the glyphosate and Ethomeen formulation, herbicide residue (detected as phosphorous, Fig. 11b) is clearly present in the anticlinal wall regions; however, the secondary electron mode image (Fig. 11a) clearly shows a partitioning of the material in the anticlinal wall regions. While a part of the negative charge associated with glyphosate should be neutralized by electrostatic interaction with Ethomeen, far more glyphosate is present on a molar basis than can be accommodated by the Ethomeen. Two species should predominate in the residue: one mostly glyphosate and the other a mixture of glyphosate and Ethomeen. Figure 11 can be interpreted as showing just such a partitioning.

Nonionic surfactants present a strikingly different picture. In general, globules of herbicide residue are found in the anticlinal cell wall regions. The dicamba and Triton-X100 formulation forms fairly large globules of residue (Fig. 12a). The corresponding X-ray mode image is shown in Fig. 12b. The glyphosate and Triton X-100 formulation is more complex: globules of herbicide (detected as phosphorous, Fig. 13) are seen residing in a matrix of presumed surfactant.

The nonionic surfactants have both hydrophobic and hydrophilic domains and should interact to some extent with both dicamba and glyphosate. With both herbicides this yields a mixture with a net negative charge, although the glyphosate formulation is likely to be more negatively charged. Herbicide associated with nonpolar material should be able to dissolve into epicuticular material, leaving globules of herbicide not associated with surfactant behind.

Glyphosate formulations using freshly prepared Silwet L-77 and old Silwet L-77 produce a different residue pattern. Fresh Silwet L-77 is an excellent spreader and effects stomatal infiltration (Buick et al. 1992). Little or no herbicide residue is found when it is used (Fig. 14). Older Silwet L-77 solutions undergo changes in acidic environments (Knoche et al. 1991) which make them poor spreaders and unable to effect

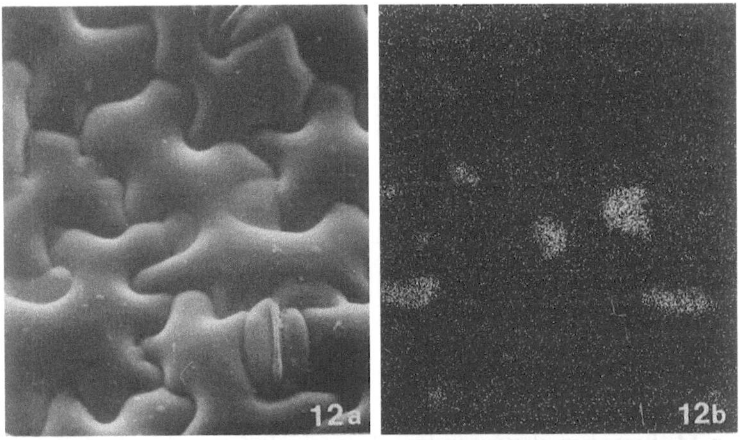

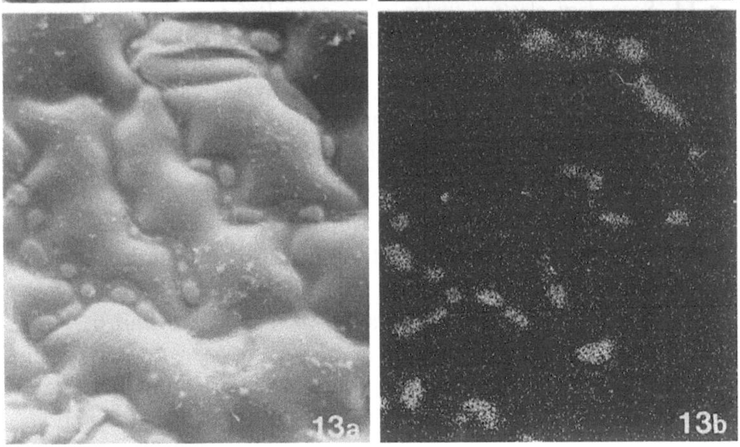

Fig. 12. a Secondary electron image of *V. faba* leaf treated with dicamba + Triton X-100, × 700. **b** Chlorine X-ray image of **a**. (From Falk and Schulke 1993)

Fig. 13. a Secondary electron image of *V. faba* leaf treated with glyphosate + Triton X-100, × 700. **b** Phosphorous X-ray image of **a**. (From Falk and Schulke 1993)

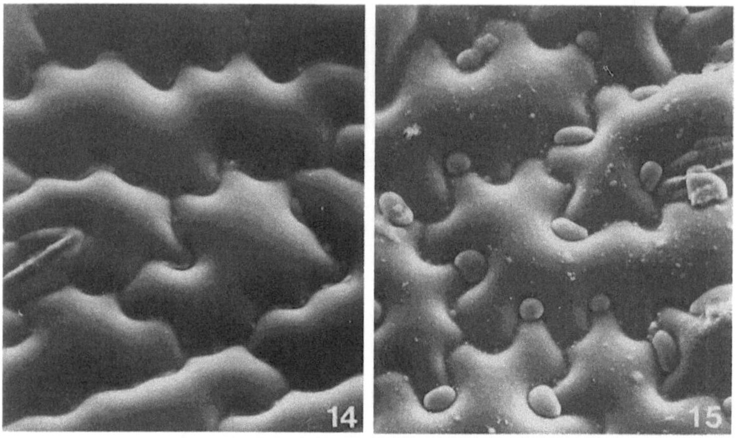

Fig. 14. Secondary electron image of *V. faba* leaf treated with glyphosate + freshly prepared Silwet L-77, × 700. (From Falk and Schulke 1993)

Fig. 15. Secondary electron image of *V. faba* leaf treated with glyphosate + old Silwet L-77, × 700. (From Falk and Schulke 1993)

stomatal infiltration. Figure 15 shows the distinct globular residues that these older Silwet L-77 formulations produce.

The herbicide formulations used were applied as 1-μl droplets, and no runoff occurred. The absolute amounts of herbicide applied are readily detected by the methods used. An inability to find herbicide residue on a sample or a diminution in the amount present when compared to herbicide only treatments suggests that some amount of herbicide has been taken up. Bioassay studies not presented here support the scanning electron microscope/X-ray microanalysis study.

References

Amelunxen F, Morgenroth K, Picksak T (1967) Untersuchungen an der Epidermis mit dem Stereoscan-Elektronenmikroskop. Z Pflanzenphysiol 57:79–95

Anderson NH, Girling J (1983) The uptake of surfactants into wheat. Pestic Sci 14:399-404

Baker EA (1982) Chemistry and morphology of plant epicuticular waxes. In: Cutler DF, Alvin KL, Price CE (eds) The plant cuticle. Academic, New York, pp 139–165

Baker EA, Hunt GM (1988) Factors affecting foliar penetration and translocation pesticides. In: Cross B, Scher HB (eds) Pesticide formulations: innovations and developments. American Chemical Society, Washington DC, pp 8–21 (ACS symposium series 371)

Baker EA, Hayes AL, Butler RC (1992) Physiochemical properties of agrochemicals: their effects on foliar penetration. Pestic Sci 34:167–182

Benzing DQ, Burt KM (1970) Foliar permeability among twenty species of the Boroeliaceae. Bull Torrey Bot Club 97:269–279

Bianchi G (1985) Epicuticular wax analysis in Triticum and related species. Genet Agr 39:471–486

Bukovac MJ, Whitmoyer RE, Reichard DR (1983) Nature and effects of surfactant deposits on leaf surfaces. Abstr Hort Sci 18:618–619

Buick RD, Field RJ, Robson AB, Buchan GD (1992) A foliar uptake model of triclopyr. In: Foy CL (ed) Adjuvants for agrochemicals. CRC Press, Boca Raton, pp 87–99

Chamel A (1986) Foliar absorption of herbicides: study of the cuticular penetration using isolated cuticles. Physiol Veg 24:491–508

Crafts AS (1961) The chemistry and mode of action of herbicides. Interscience, New York

de Ruiter H, Verbeek MAM, Uffing AJM (1988) Mode of action of a nonionic and a cationic surfactant in relation to glyphosate. In: Cross B, Scher HB (eds) Pesticide formulations: innovations and developments. American Chemical Society, Washington DC, pp 44–55 (ACS symposium series 371)

Falk RH, Schulke G (1993) Partition patterns of surfactant-herbicide mixtures on plant surfaces using scanning electron microscopy and X-ray microanalysis. J Pestic Sci (in review)

Falk RH, Bayer DE, Hess FD (1989) Structure-activity-residue relationships of alkylphenol polyoxyethylene non-ionic surfactants and MSMA. Pestic Sci 27:243–251

Falk RH, Guggenheim R, Schulke G (1993) Surfactant-induced phytotoxicity. Weed Technol (in press)

Field RJ, Bishop NG (1988) Promotion of stomatal infiltration of glyphosate by the organosilicone surfactant reduces the critical rainfall period. J Pestic Sci 24:55–62

Greene DW, Bukovac MJ (1974) Stomatal penetration: effect of surfactants and role in foliar absorption. Am J Bot 61:100–106

Gülz P-G, Scora RW, Müller E, Marner F-J (1987) Epicuticular leaf waxes of Citrus halimii stone. J Agric Food Chem 35:716–720

Hallam ND, Juniper BE (1971) The anatomy of the leaf surface. In: Preece TF, Dickinson CH (eds) Ecology of leaf surface micro-organisms. Academic, New York, pp 3–37

Harr J, Guggenheim R, Schulke G, Falk RH (1991) The leaf surface of major weeds. Sandoz, Witterswil

Hart CA, Young BW (1987) Scanning electron microscopy and cathodolujinescence in the study of interactions between spray droplets and leaf surfaces. Aspects Appl Biol 14:127–40

Hartley GS, Graham-Bryce IJ (1980) Physical properties of pesticide behavior, vol 1. Academic, New York

Hess FD, Bayer DE, Falk RH (1974) Herbicide dispersal patterns. I. As a function of leaf surface. Weed Sci 22:394–401

Hess FD, Falk RH, Bayer DB (1975) Herbicide dispersal patterns. II. Mapping residues using x-ray fluorescence. Weed Sci 23:308-314

Hess FD, Bayer DE, Falk RH (1981) Herbicide dispersal patterns. III. As a function of formulation. Weed Sci 29:224–229

Hess FD, Goss JR, Bucholtz DL, Falk RH (1987) The physical form of flamprop-ethyl herbicide on sprayed leaves influences absorption and subsequent efficacy. In: Greenhalgh R, Roberts TR (eds) Pesticide science and biotechnology. Blackwell, London, pp. 209–214

Holloway PJ (1982) The chemical constitution of plant cutins. In: Cutler DF, Alvin KL, Price CE (eds) The plant cuticle. Academic, New York, pp 45–85

Holloway PJ (1993) Structure and chemistry of plant cuticles. Pestic Sci 37:203–232

Hull HM, Morton HL, Wharrie JR (1975) Environmental influences on cuticle development and resultant foliar penetration. Bot Rev 41:421–452

Jyung WH, Wittwer SH, Bukovac MJ (1965) The role of stomata in the foliar absorption of Rb by leaves of tobacco, bean and tomato. Proc Am Soc Hort Sci 86:361–367

Knoche M, Bukovac MJ (1991a) Studies on octylphenoxy surfactants. IX. Effect of oxyethylene chain length on GA3 absorption by sour cherry leaves. J Plant Growth Regul 10:173–177

Knoche M, Tamura H, Bukovac MJ (1991b) Performance and stability of the organosilicone surfactant Silwet L-77: effect of pH, concentration and temperature. J Agric Food Chem 39:202–206

Knoche M, Noga G, Lenz F (1992) Surfactant-induced phytotoxicity: evidence for interaction with epicuticular wax fine structure. Crop Protect 11:51–56

Kolattukudy PE (1980) Biopolyester membranes of plants: cutin and suberin. Science 208:990–1000

Kolattukudy PE (1981) Structure, biosynthesis, and biodegradation of cutin and suberin. Annu Rev Plant Physiol 32:539–567

Kuzych IJ, Meggitt WF (1983) Alteration of epicuticular wax structure by surfactants. Micron Microsc Acta 14:279–280

MacIsaac SA, Paul RN, Devine MD (1991) A scanning electron microscope study of glyphosate deposits in relation to foliar uptake. Pestic Sci 31:53–64
Martin JT, Juniper BE (1970) The cuticles of plants. Arnold, London
Miller RH (1985) The prevalence of pores and canals in leaf cuticular membranes. Ann Bot 55:459–471
Miller RH (1986) The prevalence of pores and canals in leaf cuticular membranes. II. Supplemental studies. Ann Bot 57:419–434
Noga G, Wolter M, Barthlott W, Petry W (1991) Quantitative evaluation of epicuticular wax alterations as induced by surfactant treatment. Angew Bot 65:239–252
Noga GJ, Knoche M, Wolter M, Barthlott W (1987) Changes in leaf micromorphology induced by surfactant application. Angew Bot 61:521–528
Ormrod DJ, Renney AJ (1968) A survey of weed leaf stomata and trichomes. Can J Plant Sci 48:197–209
Parr JF, Norman AG (1965) Considerations in the use of surfactants in plant systems: a review. Bot Gaz 126:86–96
Price CE (1982) A review of the factors influencing the penetration of pesticides through plant leaves. In: Cutler DF, Alvin KL, Price CE (eds) The plant cuticle. Academic, New York, pp 237–252
Rawlinson CJ, Muthyalu G, Turner RH (1978) Effect of herbicides on epicuticular wax of winter oilseed rape (Brassica napus) and infection by Pyrenopeziza brassicae. Trans Br Mycol. Soc 71:441–451
Richard EP, Slife FW (1979) In vivo and in vitro characterization of the foliar entry of glyphosate in hemp dogbane (Apocynum cannabinum). Weed Sci 27:426–433
Riederer M, Schönherr J (1984) Accumulation and transport of (2,4-dichlorophenoxy)acetic acid in plant cuticles. I. Sorption in the cuticular membrane and its components. Ecotoxicol Environ Safety 8:236–247
Sargent JA (1965) The penetration of growth regulators into leaves. Annu Rev Plant Physiol 16:1–12
Schönherr J, Bukovac MJ (1970) Preferential polar pathways in the cuticle and their relationship to ectodesmata. Planta 92:189–201
Schönherr J, Bukovac MJ (1972) Penetration of stomata by liquids: dependence of surface tension, wettability and stomatal morphology. Plant Physiol 49:813–819
Schönherr J, Bukovac MJ (1973) Ion exchange properties of isolated tomato fruit cuticular membrane: exchange capacity, nature of fixed charges and cation selectivity. Planta 109:73–93
Schönherr J, Huber R (1977) Plant cuticles are polyelectrolytes with isoelectric points around three. Plant Physiol 59:145–150

Silcox D, Holloway PJ (1989) Foliar absorption of some nonionic surfactants from aqueous solution in the absence and presence of pesticidal active ingredients. In: Chow PNP, Grant CA, Hinshalwood AM, Simundsson E (eds) Adjuvants and agrochemicals, vol 1. CRC Press, Boca Raton, pp 115–128

Steurbaut W, Melkebeke G, DeJonckheere W (1989) The influence of nonionic surfactants on the penetration and transport of systemic fungicides in plants. In: Chow PNP, Grant CA, Hinshalwood AM, Simundsson E (eds) Adjuvants and agrochemicals, vol 1. CRC, Boca Raton, pp 93–103

Stevens PJG (1984) Molecular properties of the penetrant and other physicochemical factors affecting foliar uptake and translocation of pesticides from spray droplets. PhD thesis, Bristol University

Stevens PJG, Baker EA (1987) Factors affecting the foliar absorption and redistribution of pesticides. I. Properties of leaf surfaces and their interactions with spray droplets. Pestic Sci 19:265–281

Stevens PJG, Bukovac MJ (1987a) Studies on octylphenoxy surfactants. I. Effects of oxyethylene content on properties of potential relevance to foliar absorption. Pestic Sci 20:19–35

Stevens PJG, Bukovac MJ (1987b) Studies on octylphenoxy surfactants. II. Effects on foliar uptake and translocation. Pestic Sci 20:37–52

Stevens PJG, Gaskin RE, Sung-Ok H, Zabkiewicz JA (1992) Pathways and mechanisms of foliar uptake as influenced by surfactants. In: Foy CL (ed) Adjuvants for agrochemicals. CRC Press, Boca Raton, pp 385–398

Takeno T, Foy CL (1974) Effect of polysorbate surfactants on the ultrastructure of leaf surfaces. Abstr Weed Sci Soc Am 27:92

Tulloch AP, Hoffman LL (1973) Leaf wax of oats. Lipids 8:617–622

Whitehouse P, Holloway PJ, Caseley JC (1982) The epicuticular wax of wild oats in relation to foliar entry of the herbicides diclofop-methyl and difenzoquat. In: Cutler DF, Alvin KL, Price CE (eds) The plant cuticle. Academic, London, pp 315-330

Wilkinson RE, Mayeux HS (1987) Composition of epicuticular wax on Isocoma leaves. Bot Gaz 148:12–16

Wolter M, Barthlott W, Knoche M, Noga GJ (1988) Concentration effects and regeneration of epicuticular waxes after treatment with Triton X-100 surfactant. Angew Bot 62:53–62

Wyrill JB, Burnside OC (1977) Glyphosate toxicity to common milkweed and hemp dogbane as influenced by surfactants. Weed Sci 25:275–287

Zabkiewicz JA, Coupland D, Ede F (1988) Effects of surfactants on droplet spreading and drying rates in relation to foliar uptake. In: Cross B, Scher HB (eds) Pesticide formulations: innovations and developments. American Chemical Society, Washington DC, pp 77–89 (ACS symposium series 371)

4 Physicochemical Factors Influencing the Adjuvant-Enhanced Spray Deposition and Coverage of Foliage-Applied Agrochemicals

P. J. Holloway

4.1 Introduction

Most pesticides are applied to foliage as sprays of water-based formulations atomised through hydraulic nozzles. It is well known that this method of transfer is inherently inefficient, resulting in only a small proportion of the dose applied actually being deposited on the intended target, with even less eventually reaching the ultimate site of biological action. In the crucial first stage of the process, from atomisation to droplet retention and deposit formation, a number of factors contribute to the shortfall, the most important of which are drift, in-flight evaporation and, especially, reflection and run-off from the target (Young 1986; Hall 1990; Hall et al. 1993). There are now increasing environmental and regulatory pressures on manufacturers to seek ways of alleviating these fundamental problems of crop spraying and, at the same time, reducing pesticide doses.

Two main approaches are currently being explored to make spray application more efficient, one involving redesign of the spray machinery itself, the other based on the use of appropriate built-in or tank-mix adjuvants that modify the properties and behaviour of spray droplets of the pesticide formulation at field dilution. On the purely technological side it has already been shown, for example, that air assistance or induction of an electrostatic charge during delivery may result in as much as 50% reduction in spray drift with better retention; improved nozzle designs or conformations may be of benefit or even essential in ensuring efficient application of smaller spray volumes (reviewed by Göhlich 1992; Miller 1992). Such aspects, although complementary, lie beyond the scope of the present article which focuses on the alternative formulation approach using conventional spray equipment. It examines the physicochemical interactions between adjuvants, spray droplets and plant surfaces which appear to be most conducive to enhanced spray retention and target coverage. Illustrations are provided by some recent data obtained at the Long Ashton laboratory.

4.2 Factors Influencing Retention of Spray Droplets

The physical behaviour of water-containing droplets when they impact with a solid surface has been documented by high-speed photography and the energy transfer processes involved described in detail (Ford and Furmidge 1967; Lake 1977; Young 1980; Spillman 1984; Tadros 1987; Wirth et al. 1991; Brazee et al. 1991). The chain of possible events is summarised in Fig. 1. When droplets from a hydraulic nozzle are sprayed onto leaves, three things may happen depending mainly on the droplet size, the droplet velocity and the intrinsic physicochemical properties of both droplet and leaf surface. Thus, droplets may be retained, reflected or shatter into smaller satellites which may be lost or undergo further impaction.

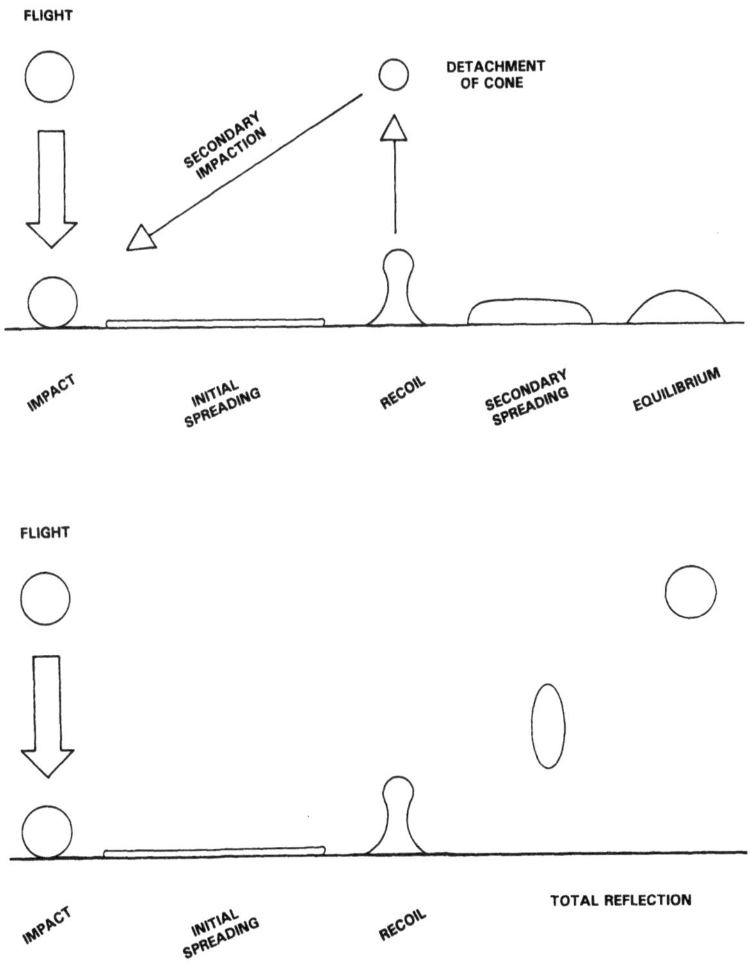

Fig. 1. Summary of the possible impaction behaviour of spray droplets on a leaf surface. On difficult-to-wet targets, most droplets ca. 100 μm in diameter are likely to be retained; droplets larger than 300 μm in diameter probably shatter or are totally reflected from the surface unless an efficient spray modifier is added to the spray liquid. (Adapted from Ford and Furmidge 1967; Young 1980; Wirth et al. 1991)

Droplet reflection is obviously the most serious problem for spray application. It may occur from either a dry or a uniformly wetted leaf surface and is caused by an air film which becomes trapped between the impacting droplet and the surface (Hartley 1967). When this occurs on dry leaf surfaces, it can always be attributed to the presence of microroughness, usually wax crystals or a dense covering of trichomes. The phenomenon of droplet reflection was examined in detail more than 30 years ago in the elegant studies of Brunskill (1956) and Hartley and Brunskill (1958) on the waxy and highly water-repellent surface of pea (*Pisum sativum*) leaves. Using a droplet generator capable of producing monosize droplets in the range 80–500 µm in diameter and aqueous solutions of organic solvents of known surface tension, they demonstrated that there is not only a critical droplet diameter but also a critical surface tension, above or below which retention is either very low or very high, respectively. The critical diameter for droplets of pure water (72 mN/m) on pea was ca. 100 µm, but this could be increased to ca. 250 µm by lowering the surface tension of the droplet to ca. 45 mN/m. The sigmoidal relationship between retention and surface tension was also confirmed later by Anderson and Hall (1989) for 200-µm droplets of aqueous methanol applied to leaves of the same species. A multiplicity of interactions is therefore likely to occur when the mixed spectrum of droplets from a hydraulic nozzle, which typically lie in the range of 10–500 µm in diameter, hits a plant surface.

The results obtained from model experiments such as those described above, however, do provide a simple rationale for optimising spray retention – reduce droplet size and/or the surface tension of impacting droplets. The influence of some typical adjuvants on foliar retention and their relationship with these spray parameters is examined in the following sections.

4.2.1 Adjuvant Effects on Spray Retention

The bulk of the adjuvant market is dominated by nonionic surfactants, many of which are sold specifically as wetters or deposition agents (Holloway 1993). Nevertheless, comparatively little quantitative information is available to show just how effective they are at increasing spray retention on different plant species. De Ruiter et al. (1990) have compared the track-sprayer performance of a nonylphenol with a mean molar ethylene oxide content (EO) of 8, with a 15 EO tallowamine at various concentrations on six different species using a spray volume of 230 l/ha. They found that the two surfactants produce similar effects on the retention of fluorescein which is not enhanced compared with water on the wettable leaves of black nightshade (*Solanum nigrum*), tomato (*Lycopersicon esculentum*) and chamomile (*Chamomilla recutita*). On the water-repellent species examined, wheat (*Triticum aestivum*), couch grass (*Elymus repens*) and pea, however, a linear relationship between retention and logarithm of surfactant concentration was observed, with up to a 20-fold increase using 10 g/l surfactant. Concentration-dependent adjuvant enhancement of the retention of tebuconazole on maize (*Zea mays*) was also quantified by Wirth et al. (1991) when spraying at 250 l/ha. Irrespective of concentration, much better retention was obtained with a 6 EO trimethylnonanol than with a 6 EO tridecanol or an ethoxylated organosilicone; maximum retention was achieved with the surfactants at 1 g/l. However, the performance of two high molecular weight polyvinyl alcohols was equivalent to that of the best surfactant.

In our research at Long Ashton, we have made more comprehensive surveys of different spray additives under identical track sprayer conditions to examine possible structure–retention enhancement relationships. An even-spray hydraulic nozzle delivering ca. 200 l/ha dilute aqueous solution of fluorescein was used throughout. For conventional surfactants, such as those based on primary alcohol or nonylphenol hydrophobes, EO content was found to be an important factor influencing the efficiency of spray capture on difficult-to-wet targets such as oats (*Avena sativa*), rape (*Brassica napus*) and brome (*Bromus sterilis*). Products with EO contents less than 10 performed poorly, even at high concentrations; however, there was usually little difference between the retention-enhancing properties of those of higher EO content (10–20) which increased with concentration (Fig. 2). Similar effects were not observed on more wettable species, such as field bean (*Vicia faba*), sugar beet (*Beta vulgaris*) and cleavers (*Galium aparine*), where the retention of surfactant-containing solutions of fluorescein does not significantly differ from that of water.

Further investigations were carried out to directly compare the relative efficiencies of commonly used adjuvants on spray deposition on the same target species. The results obtained from one such experiment on

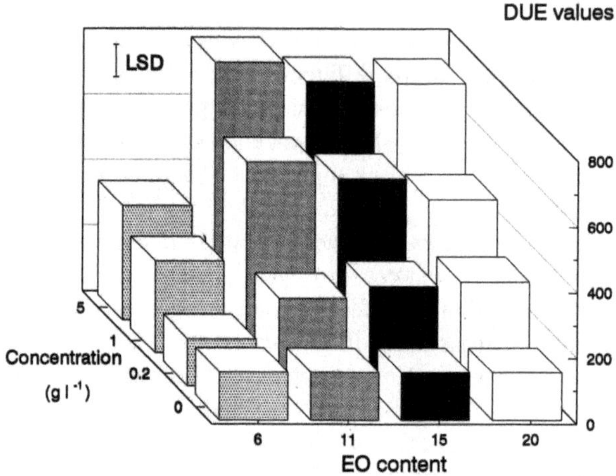

Fig. 2. Effect of ethylene oxide (*EO*) content and concentration of C_{13}/C_{14} alcohol surfactants on the retention of 0.1 g/l aqueous solutions of fluorescein when sprayed at 160 l/ha onto the foliage of oat plants. *DUE*, Deposit per unit emission (ng g^{-1} dry weight g^{-1} fluorescein applied ha^{-1}); *LSD*, $p = 0.05$. (Adapted from van Toor et al. 1993)

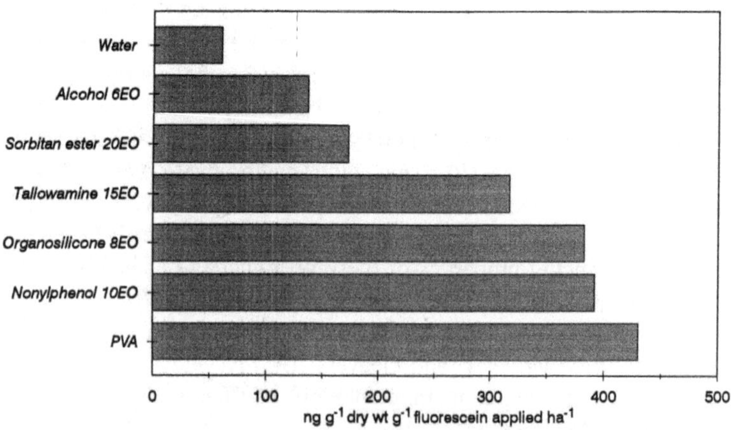

Fig. 3. Effect of some surfactants at 1 g/l on the retention of 0.1 g/l aqueous solutions of fluorescein when sprayed at 160 l/ha onto the foliage of rape plants. (Adapted from Holloway et al. 1993)

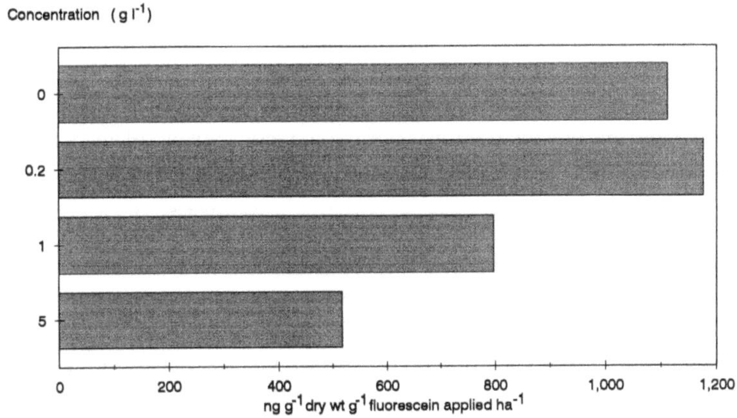

Fig. 4. Effect of concentration of an 8 EO organosilicone surfactant on the retention of 0.1 g/l aqueous solutions of fluorescein when sprayed at 160 l/ha onto the foliage of sugar beet plants. (Adapted from Holloway et al. 1993)

rape are shown in Fig. 3. At 1 g/l retention was improved by a variety of surfactant types, the best being obtained with tallowamine 15 EO, organosilicone 8 EO and nonylphenol 10 EO. However, the maximum effect on this target was produced by inclusion of a polyvinyl alcohol (M_r 49 000), thus confirming the findings of Wirth et al. (1991) for this class of adjuvant.

On more easily wetted species, none of the adjuvants that we tested in this way were found to improve retention performance substantially when compared with water alone. However, when using an 8 EO organosilicone, reduced spray retention was often observed, the magnitude of which increased with surfactant concentration (Fig. 4). Because such surfactants have exceptionally high surface activity (Sect. 4.3.1), decreased performance can probably be attributed to droplet coalescence and run-off. Droplet reflection from the film of liquid which rapidly forms on the leaf surface may also be a contributing factor.

Simple quantitative foliage retention measurements with solutions of dyes can be useful indicators of the potential of an adjuvant as a spray modifier. However, their effectiveness must be confirmed ultimately in the presence of the pesticide formulation being used. It is often assumed that foliar retention is increased simply by adding a surfactant to a spray solution. This is clearly not the case; much depends on its chemical composition, its concentration and the plant species being sprayed. When designing formulations, it should also be borne in mind that surfactants are not the only class of compounds capable of increasing spray deposition.

4.2.2 Relationships with Surface Tension

The droplet generator experiments with aqueous solutions of organic solvents have highlighted the pivotal role that surface tension plays in droplet retention on a water-repellent surface. In such solutions, however, a reduction in surface tension is produced by altering the bulk properties of water and, thus, surface equilibrium is achieved instantaneously (Anderson and Hall 1989). Consequently, the surface tension of any spray droplet changes little during and after impact. In surfactant-containing solutions, lowering of surface tension is caused by the adsorption of surfactant molecules at the air-water interface; this process is diffusion-controlled, and therefore time-dependent. The increase in surface tension which occurs at any nonequilibrium surface age is known as the dynamic surface tension and is highly relevant to the impaction and retention process. For droplets from a hydraulic nozzle it has been estimated that the time required for atomisation is 3–5 ms, for flight anywhere between 30 and 400 ms and for flattening and recoil less that 1 ms (Reichard 1988; Anderson and Hall 1989; Wirth et al. 1991). Within the time-frame of any of these events (Fig. 1) it is unlikely that the surface of surfactant-containing droplets is at equilibrium; the surface tension changes continuously at the air-water interface and also at the water-leaf surface interface during the later stages. It is hardly surprising therefore that the equilibrium surface tension values of surfactant solutions are not correlated with their foliar retentivity. Not only are the concentration-dependent effects unexplained, but also our observation that surfactants with markedly different equilibrium values, for example, tallowamine 15 EO (40 mN/m), nonylphenol 10 EO (30 mN/m) and organosilicone 8 EO (22 mN/m), can enhance retention to a remarkably similar degree under hydraulic spraying conditions (Fig. 3). The latter provides circumstantial evidence that factors other than surface

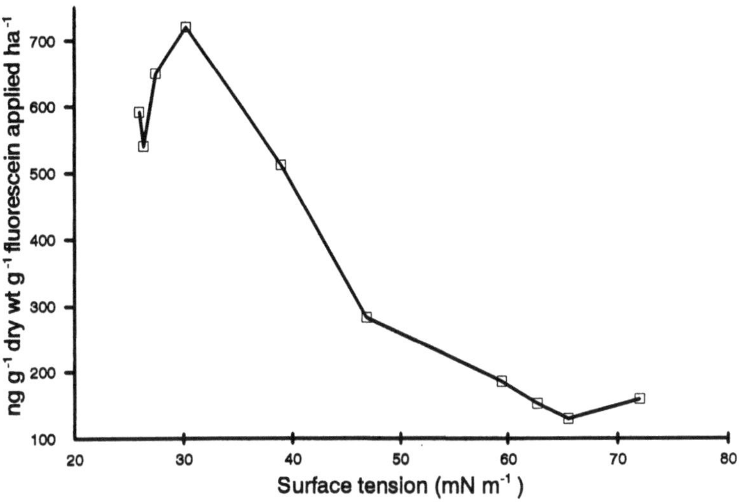

Fig. 5. Influence of the surface tension of aqueous solutions of *n*-propanol on the retention of 0.1 g/l aqueous solutions of fluorescein when sprayed at 160 l/ha onto the foliage of oat plants. The propanol contents of the solutions were 0, 1, 5, 10, 50, 100, 200, 300, 400 and 500 ml/l, respectively. (Adapted from Holloway et al. 1993)

tension may also be contributing to efficient spray deposition in these circumstances.

Several workers have reported that droplet retention in the presence of a surfactant is significantly correlated with the dynamic surface tension of the spray solution (Anderson and Hall 1989; de Ruiter et al. 1990; Wirth et al. 1991; Stevens et al. 1993) and a predictive model has been developed and tested (Grayson et al. 1991, 1993). Although a variety of methods have been used to measure surface tension, the relationship appears to be valid over a wide range of nonequilibrium surface ages, not all of which would be appropriate to those of impacting droplets. Our data for aqueous solutions of *n*-propanol sprayed onto oats (Fig. 5) give some indication of the actual surface tension values that are necessary at impaction in ensure maximum foliar retention on a difficult-to-wet target. Remember that for such solutions there are no dynamic surface tension effects.

We have also examined the dynamic behaviour of solutions of several different surfactants in relation to their foliar retention and sometimes found inconsistencies between the two properties. The dynamic surface tension profiles, measured by the maximum bubble pressure method, for the four C_{13}/C_{14} alcohol surfactants used to obtain the retention data given in Fig. 2, are shown in Fig. 6. They provide a good illustration of the considerable influence that concentration has on the dynamic stability of a surfactant solution; this can be simply assessed by inspection of the rate of increase of surface tension from equilibrium with increasing bubble frequency, i.e. decreasing surface age. In the case of the 6 EO product, inferior retention performance is clearly associated with poor dynamic properties even at high concentrations. It should be noted here that surfactants with such EO contents often have low equilibrium surface tensions and low water solubility, forming water-dispersible opaque solutions. The three higher EO products produced much more dynamically stable solutions, and their retention qualities, accordingly, were improved. Within this series therefore a good correlation was found between retention and the dynamic surface tension at 5 Hz (Fig. 2, Table 1). However, when comparing different surfactant types, absolute dynamic surface tension values were less reliable predictors of their relative retention qualities. For example, on rape, although the retention of solutions of nonylphenol 10 EO, tallowamine 15 EO, sorbitan monolaurate 20 EO and organosilicone 8 EO increased as dynamic surface tension decreased, the amounts of retention observed with solutions of similar surface tension could differ (Table 2). Although at surface tensions greater than 60 mN/m retention was poor in all cases, good retention could be obtained from solutions with values as high as 54 (tallowamine 15 EO, sorbitan monolaurate 20 EO) or as low as 39 (nonylphenol 10 EO). The consistently lower dynamic surface tensions of organosilicone 8 EO solutions did not ensure that there was a concomitant increase in spray capture on oats. Their general performance was no better than that of the more conventional surfactants, as we have also found on several other waxy species. Close scrutiny of the data of de Ruiter et al. (1990) also reveals a

Fig. 6. Dynamic surface tension profiles of aqueous solutions of four C_{13}/C_{14} → alcohol surfactants at 0.2, 1 and 5 g/l. Values determined by the maximum bubble pressure method using the Krüss Dynamic Surface Tensionmeter BP–10 (Holloway, unpublished).

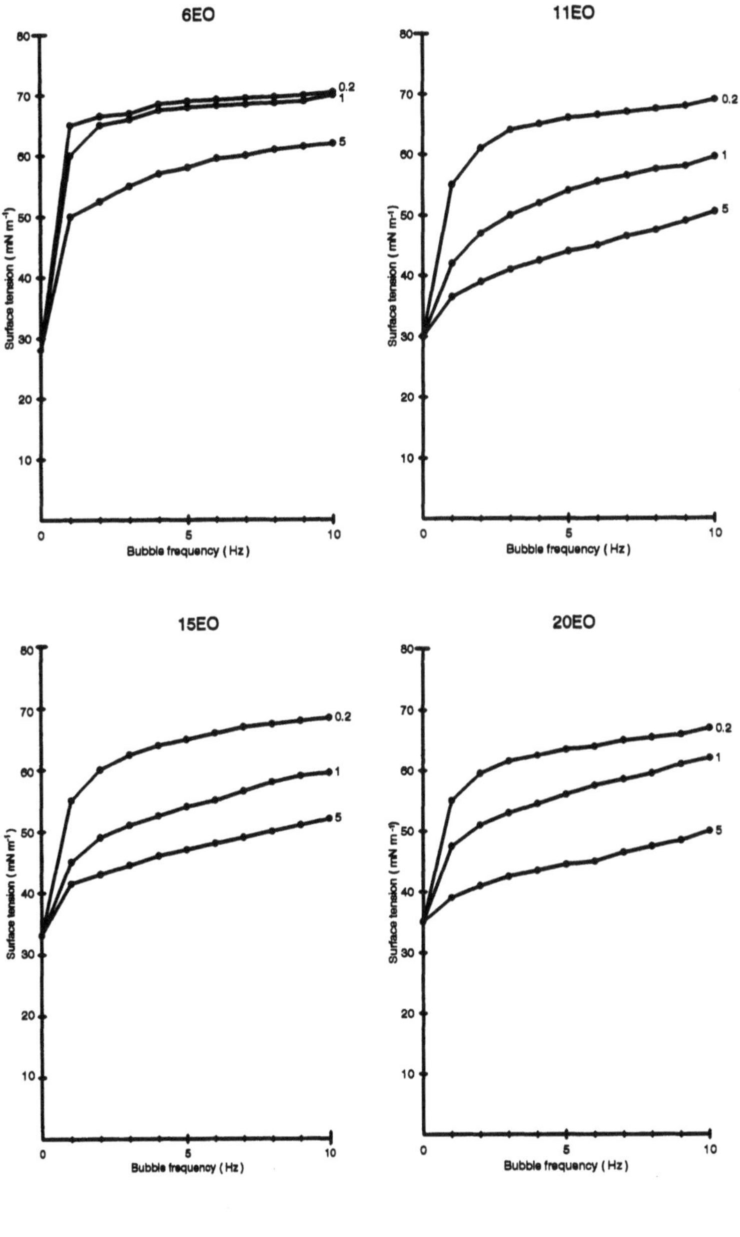
6EO
11EO
15EO
20EO
Surface tension (mN m⁻¹)
Bubble frequency (Hz)
0.2
1
5

Table 1. Maximum bubble pressure dynamic surface tension values (mN/m) of four C_{13}/C_{14} alcohol surfactants at a bubble frequency of 5 Hz

	Concentration (g/l)		
EO content	0.2	1	5
6	69	68	58
11	66	54	44
15	65	54	47
20	64	56	44

Table 2. Dynamic surface tension of solutions of four surfactants and their retention when sprayed through an even-spray hydraulic nozzle onto the foliage of rape at 160 l/ha (adapted from Holloway et al. 1993)

Surfactant added (g/l)	DUE[a]	Dynamic surface tension[b]
None	51	72
Nonylphenol 10 EO (0.2)	157	65
Nonylphenol 10 EO (1)	392	52
Nonylphenol 10 EO (5)	454	39
Tallowamine 15 EO (0.2)	182	66
Tallowamine 15 EO (1)	317	59
Tallowamine 15 EO (5)	485	54
Sorbitan monolaurate 20 EO (0.2)	91	66
Sorbitan monolaurate 20 EO (1)	173	60
Sorbitan monolaurate 20 EO (5)	514	54
Organosilicone 8 EO (0.2)	124	63
Organosilicone 8 EO (1)	384	47
Organosilicone 8 EO (5)	363	38

[a] Deposit per unit emission (ng g^{-1} wt foliage g^{-1} fluorescein applied ha^{-1}.
[b] Maximum bubble pressure method at 5 Hz.

discrepancy between the magnitude of enhanced retention on wheat and pea, and dynamic surface tension, a point not mentioned by the authors.

The apparently anomalous retention enhancement behaviour of some surfactant solutions indicates the likely involvement of other physicochemical properties in addition to the reduction of surface tension at the droplet surface. Surface elasticity is one possible factor, and a striking illustration of its influence is provided by solutions of polyvinylalcohols. Although these polymers have a very low surface activity (> 60 mN/m), they can enhance retention on difficult-to-wet targets to almost the same degree as the best surfactants available at equivalent concentrations in the spray solution. Polyvinylalcohols form large loops and tails in solution, and it is thought that the presence of these structures enhances the elasticity of impacting droplets (Wirth et al. 1991), thus minimising the likelihood of bounce from the surface. Crease et al. (1991) have also reported that droplet bounce from cabbage (*Brassica oleracea*) could be eliminated completely by adding a linear alkyl epoxide polymer to the spray formulation; this adjuvant, too, had weak surface-active properties.

At the concentrations necessary to enhance retention, surfactants also exist in solution as macromolecular structures known as micelles, a notable exception being organosilicones (Stevens et al. 1993). Although having a more compact shape than molecules of polyvinylalcohol and a lifetime shorter than 1 ms (Tadros 1987), the weights, and thus the sizes, of these aggregates are known to vary considerably according to surfactant composition (Attwood and Florence 1983); the molecular weights can exceed even those of polyvinylalcohols, reaching in excess of 1 million in some cases. It is possible that the elastic or viscoelastic properties of micelles are playing a part in droplet impaction, and this may compensate for the deficiencies in the dynamic surface tension properties of some surfactants.

Stevens et al. (1993) have suggested that dynamic surface viscosity is the factor missing in providing a satisfactory explanation for surfactant-enhanced foliar retention. However, at the concentrations involved, surfactant effects on bulk viscosity are usually small (Hall et al. 1993); any dynamic effects are even smaller. Under droplet generator conditions Brunskill (1956) found retention of glycerol solutions of constant surface tension to be relatively unaffected by viscous changes in the range of practical interest.

4.2.3 Relationships with Spray Quality

If an adjuvant has surface-active properties, these probably affect atomisation through a hydraulic nozzle because this process represents a disruption of the consolidating influence of surface tension by the action of internal and external forces (Lefebvre 1993). Using solutions of either surfactants or organic solvents, several workers have shown that the volume median diameter (VMD) of spray droplets from hydraulic pressure nozzles is directly related to concentration (Arnold 1983), and thus to dynamic surface tension (Anderson et al. 1988; Grayson et al. 1991). This effect has been demonstrated using both spatial and temporal sampling methods at short and long distances from the spray nozzle. If VMD is reduced, better retention would be expected because of the lower impaction energy of smaller-sized droplets (Anderson and Hall 1989). Atomisation effects are therefore an additional consideration in attempting to explain the reasons for the success or failure of an adjuvant to improve the retention performance of a hydraulic spray application.

In order to make realistic assessments of the direct relationships between foliar retention and atomised droplet-sizes, however, the choice of an appropriate method and position for sampling is an important consideration because the quality of the spray cloud is likely to change as the distance from the nozzle increases (Arnold 1983) and to vary at different positions along a spray fan (Grayson et al. 1991). Because deposition on plants depends on the spray generated per unit time, temporal droplet sampling should be the method of choice. With these basic criteria in mind, we measured in-flight droplet-size spectra with a phase Doppler particle analyser under conditions which simulate those used in all of our track sprayer retention experiments utilising an even-spray nozzle. Sampling the middle 16 cm of the spray cloud at a distance of 40 cm from the nozzle should thus be representative of the actual spectrum of droplets at the time of impact with the various plant targets used. Of the many droplet size parameters measured, only VMD and the percentage of the spray volume in droplets smaller that 100 μm were found to be useful indicators of the retention performance of surfactants; there was usually a strong negative correlation between these two droplet parameters. In the case of a C_{13}/C_{14} alcohol surfactant series the lack of retention-enhancing activity of the 6 EO product on oats (Fig. 2)

Table 3. Some droplet parameters for aqueous solutions of four C_{13}/C_{14} alcohol surfactants atomised through an even-spray nozzle at 200 kP_a at a flow rate of 0.44 l/min (adapted from Holloway et al. 1993)

Surfactant	Concentration (g/l)	VMD (μm)	% Volume <100 μm	Mean droplet velocity (ms^{-1})	Droplet density (ml^{-1} air volume)
None	–	220	7.9	2.0	24.6
EO 6	0.2	294	1.9	3.5	9.7
EO 6	1	288	1.9	3.6	10.1
EO 6	5	252	3.0	3.2	13.3
EO11	0.2	211	8.8	2.0	27.7
EO11	1	195	11.3	2.0	32.3
EO11	5	183	12.9	2.1	42.3
EO15	0.2	211	9.0	2.0	26.3
EO15	1	197	10.7	2.0	34.7
EO15	5	186	13.0	2.0	44.7
EO20	0,2	209	8.8	2.0	26.7
EO20	1	203	10.4	2.0	31.0
EO20	5	182	12.8	2.0	39.0

For corresponding retention data consult Fig. 2, for dynamic surface tension, Fig. 6 and Table 1.

could be related to the marked increase in VMD and consequent decrease in the small droplet component compared with water alone and the higher EO members of the same series (Table 3). Because there were fewer, smaller, more slowly moving droplets produced by atomisation of solutions of the 6 EO surfactant, mean droplet velocity was also increased and droplet density decreased. The ability of the 11, 15 and 20 EO surfactants to improve retention was associated with a decrease in VMD, especially at the highest concentration, an increase in the small droplet component, little change in mean velocity compared with water and an increase in droplet density. Patternation measurements confirmed that none of these effects were being produced by changes in the overall spray pattern.

Table 4. Droplet VMD and small droplet component of aqueous solutions of three surfactants and polyvinylalcohol (PVA) atomised through an even-spray nozzle at 200 kP_a at a flow rate of 0.44 l/min (adapted from Holloway et al. 1993)

Adjuvant	Concentration (g/l)	VMD (µm)	% Volume < 100 µm
None	–	220	7.9
Nonylphenol 10 EO	0.2	214	8.2
	1	202	10.2
	5	184	12.5
Tallowamine 15 EO	0.2	222	7.3
	1	215	9.0
	5	202	10.2
Organosilicone 8 EO	0.2	283	2.3
	1	293	2.2
	5	242	3.9
PVA (M_r 49 000)	0.2	215	7.9
	1	227	7.0
	5	242	4.8

For retention data of the three surfactants see Table 2.

Droplet-size data for three other surfactant classes and polyvinylalcohol are shown for comparison in Table 4; the corresponding retention values obtained for the former on rape are given in Table 2. Although the atomisation characteristics of the nonylphenol 10 EO and tallowamine 15 EO could be related to their retention behaviour in the same way as the higher EO C_{13}/C_{14} alcohols, those of the organosilicone 8 EO could not. The atomised droplet-size spectrum obtained from this surfactant resembled that of the 6 EO C_{13}/C_{14} (see Table 3). By analogy, retention performance would therefore have been expected to be poor, but this was not the case (Table 2). Surface tension effects probably compensate for the increase in the number of larger droplets which otherwise would be poorly retained. At the concentrations used, polyvinylalcohol had little effect on atomisation; VMD was increased slightly at the highest concentration used.

4.2.4 Predicting Deposition Performance

From the examples given in the foregoing account it is clear that it is difficult to predict the outcome of including an adjuvant on spray retention after delivery from a hydraulic nozzle. The only exception is on easy-to-wet targets, on the retention of which most of the currently-used adjuvants appear to have little or no beneficial effect compared with water alone. On less easily wetted plant surfaces the situation is more complex, and adjuvant-enhanced deposition is unlikely to depend on a single physicochemical factor but on the interaction between a number of them. At the present time, the predictive models (Anderson et al. 1988; Anderson and Hall 1989; Grayson et al. 1991) are based only on a very limited number of surfactants, one type of nozzle and one set of operating conditions, and they thus need to be validated under a wider range of circumstances to be of any real practical value. Nevertheless, they should provide a stimulus for future research.

4.3 Factors Influencing Spreading of Spray Droplets

This section is concerned with events that take place once spray droplets have been accepted by a leaf surface after impaction (Fig. 1) and have an influence on the area of leaf surface eventually covered by the pesticide when the droplet has evaporated. In practice, the droplets cover a wide range of sizes, the largest being formed either as a result of multiple impactions or by coalescence with neighbouring droplets on the surface. Coverage depends on the volume of spray applied, the structure of the target and the nature of the formulation. Maximum coverage is required for pesticides which have a contact or protectant action, but is probably of less importance for systemic ones.

Three things may happen to a spray droplet before it reaches final equilibrium on a leaf surface: it may maintain its original shape on impaction, spread or, sometimes, contract. It is well established that the major factors involved in the eventual outcome are the surface tension of the liquid and the physicochemical properties of the surface. The least amount of droplet spreading and thus contact area occurs with liquids of high surface tension on leaves which exhibit microroughness. These factors, as well as the size and speed of evaporation of the droplets, therefore contribute to the final size of the deposit area.

4.3.1 Adjuvant Effects on Droplet Spreading

The spreading properties of droplets of aqueous solutions of adjuvants on leaf surfaces can be assessed from contact angle determinations before evaporation or, after evaporation, by deposit area measurements after addition of an appropriate marker dye. Spread factors can be calculated from the results obtained by both methods (Tadros 1987) and provide a useful bench-mark for relative efficiency. Although the droplet sizes used for such measurements are invariably much larger than those of typical spray droplets, the results obtained do appear to be correlated well with the spreading behaviour observed on leaves after a hydraulic spray application.

To function as an efficient spreader, an adjuvant must possess a high degree of surface activity. The best spreaders currently available are organosilicone surfactants which give equilibrium surface tension values in aqueous solution approaching 20 mN/m; these can induce impacted spray droplets to spread to such an extent that the leaf surface becomes completely covered by a thin film of liquid. Understandably, most conventional surfactant adjuvants (equilibrium surface tension 30–40 mN/m) perform less well in comparison, and we have shown that on easily wetted leaf surfaces (contact angle of water $< 90°$) they have little or no effect on droplet spreading and deposit area (Fig. 7). Some of these surfactants, however, can improve spreading on less easily wetted leaf surfaces (contact angle of water $> 110°$), but the surface tension must be near 30 mN/m; only nonylphenol and alcohol surfactants with an EO content less than 10 satisfy these requirements (Fig. 8).

Although the equilibrium surface tension of the bulk solution at or above the critical micelle concentration indicates the spreading potential of a surfactant, it may not always be representative of that of a droplet of the same solution in contact with the leaf surface. This is clearly illustrated by the dependence of droplet spreading on surfactant concentration, especially on waxy leaf surfaces (Fig. 9). The phenomenon is almost certainly related to the adsorption of surfactant molecules at the liquid-leaf surface interface resulting in a depletion of surfactant from the

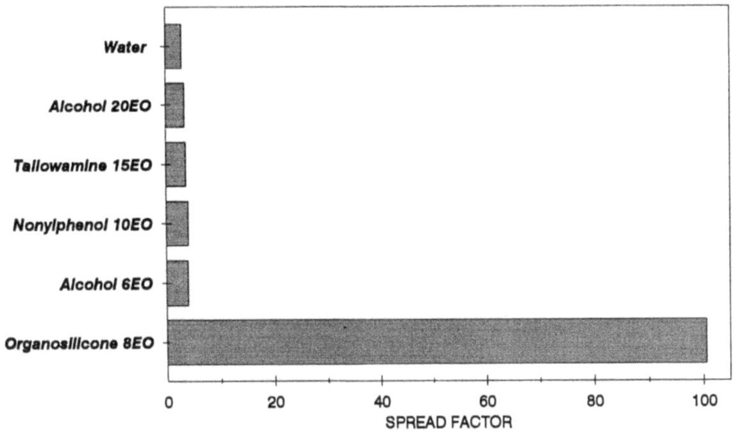

Fig. 7. Spread factors on the adaxial surface of field bean leaves for 0.2-µl droplets of water containing 0.5 g/l glyphosate, 0.5 g/l Uvitex 2B and five different surfactants at 5 g/l. The mean deposit area for water was 1.3 mm^2, that for the organosilicone 101 mm^2. (Adapted from Holloway et al. 1993)

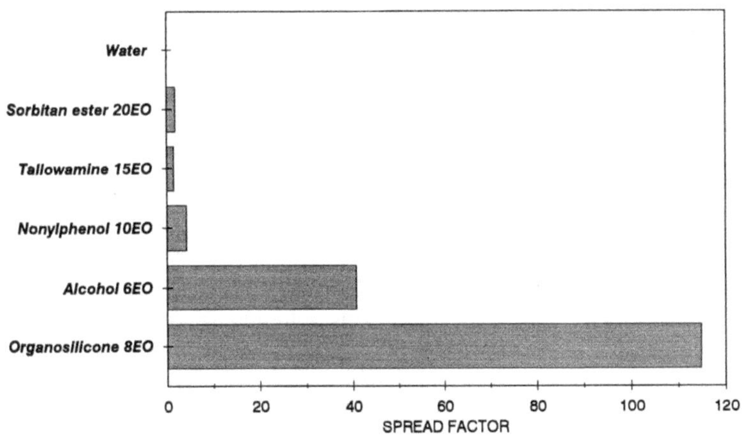

Fig. 8. Spread factors on the adaxial surface of rape leaves for 0.2-µl droplets of water containing 0.5 g/l glyphosate, 0.5 g/l Uvitex 2B and five different surfactants at 5 g/l. In the absence of surfactants, droplets were repelled by the surface. The mean deposit area for the sorbitan ester was 1.9 mm^2, that for the organosilicone 115 mm^2. (Adapted from Holloway et al. 1993)

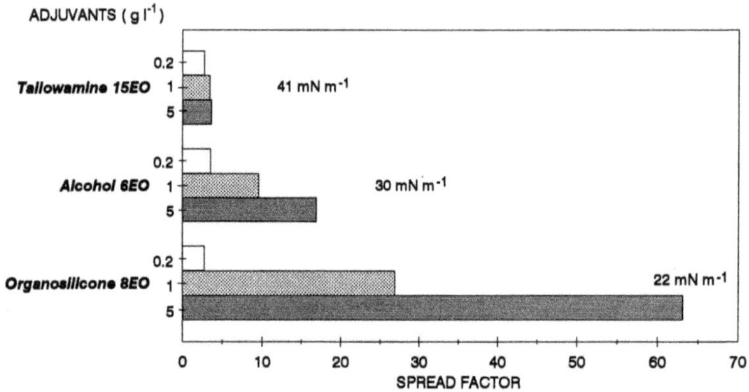

Fig. 9. Effect of concentration of three surfactants on the spread factors for 0.2-μl droplets of water containing 0.5 g/l glyphosate and 0.5 g/l Uvitex 2B on the adaxial surface of oat leaves. The mean deposit areas at 5 g/l were, 1.4 mm^2 for the tallowamine, 6.5 mm^2 for the alcohol and 63 mm^2 for the organosilicone. The corresponding equilibrium surface tension values are shown for each surfactant. (Adapted from Holloway et al. 1993)

solution, and thus causing an increase in surface tension of the droplet. At low concentrations this effect probably removes most of the surfactant from the droplet solution. Maximum droplet spreading occurs only at high surfactant concentrations, and here the surface tension, even after adsorption, should approach that of the bulk solution. Although surfactant concentration increases progressively as the droplet evaporates and may compensate for adsorption losses, this appears to be of little consequence in determining the eventual size of the spray deposit.

Spread measurements with aqueous solutions of organic solvents which are free from any adsorption effects serve to emphasise the dominant role of surface tension effects. For solutions of *n*-propanol applied to oats and rape (Fig. 10) there is a sudden increase in the ability of the droplets to spread when the surface tension reaches 25–30 mN/m; these critical surface tension values for spreading are probably the same as those required for maximum droplet retention (Sect. 4.2).

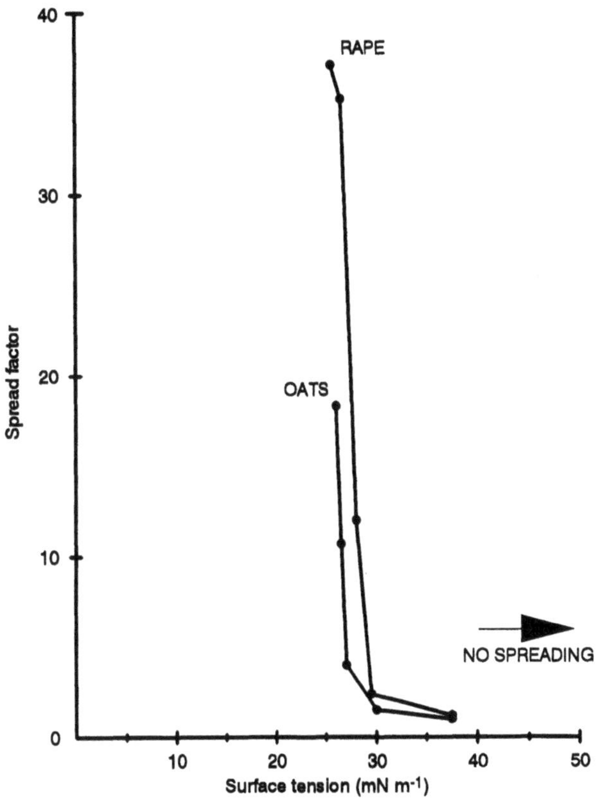

Fig. 10. Influence of surface tension on the spread factors for 0.2-µl droplets of aqueous *n*-propanol solutions containing 0.5 g/l Uvitex 2B on the adaxial surface of oat and rape leaves. The propanol contents of the solutions were 100, 200, 300, 400 and 500 ml/l, respectively. (Adapted from Holloway et al. 1993)

4.4 Conclusions

The data presented in this paper indicate that there may be some limitations to the formulation approach for optimising hydraulic spray applications using spray modifier adjuvants, such as wetters and spreaders. Such adjuvants are likely to be of benefit to spraying only on difficult-to-wet plant species. It is also clear that wetting and spreading are two distinct, and sometimes conflicting, mechanisms of spray modifier action. Enhanced deposition, that is wetting, can often be achieved without much droplet spreading, for example with high EO surfactants and polyvinylalcohols. On the other hand, although the best spreaders are always compounds with high surface activity, they may have adverse effects on spray retention. These factors should always be borne in mind for formulation strategies, especially when using surfactants as uptake activators. Compromises may have to be made to satisfy the needs of the particular pesticide formulation; adjuvant blends may be one solution to the problem.

At the present time it is not easy to predict the effect of an adjuvant on spray retention. Surface tension may not be the only factor involved, and further work is needed to identify mechanisms of enhanced droplet deposition. This is particularly important when seeking less toxic and more environmentally benign alternatives to nonylphenol and tallowamine surfactants, which are the mainstay of today's adjuvant market.

In practice, there is a maximum amount of spray retention that can be achieved on any given target before losses occur from run-off or drain-off. There is always less deposition on species where the leaves have a predominantly vertical orientation than on those which are essentially horizontal.

Acknowledgements. The author is grateful to Mrs A.L. Hayes for skilled assistance throughout, to N.M. Western for measurement of droplet spectra and to Mrs J. Hynam for production of the manuscript. The research at Long Ashton is funded by a commission from the Ministry of Agriculture, Fisheries and Food.

References

Anderson NH, Hall DJ (1989) The role of dynamic surface tension in the retention of surfactant sprays on pea plants. In: Chow PNP, Grant CA, Hinshalwood AM, Simundsson E (eds) Adjuvants and agrochemicals, vol II. CRC Press, Boca Raton, pp 51–62

Anderson NH, Hall DJ, Western NM (1988) Physicochemical aspects of spray atomisation and reduction of drop size by surfactants, vol 1. Proceedings of the International Symposium on Pesticide Application, Paris, pp 291–298

Arnold AC (1983) Comparative droplet-size spectra for three different-angled flat fan nozzles. Crop Protect 2:193–204

Attwood D, Florence AT (1983) Surfactant systems their chemistry, pharmacy and biology. Chapman and Hall, London

Brazee RD, Reichard DL, Bukovac MJ, Fox RD (1991) A partitioned energy transfer model for spray impaction on plants. J Agric Eng Res 50:11–24

Brunskill RT (1956) Physical factors affecting the retention of spray droplets on leaf surfaces. Proceedings of the 3rd British Weed Control Conference 2:593–603

Crease GJ, Hall FR, Thacker JRM (1991) Reflection of agricultural sprays from leaf surfaces. J Environ Sci Health B26:383–407

De Ruiter H, Uffing AJM, Meinen E, Prins A (1990) Influence of surfactants and plant species on leaf retention of spray solutions. Weed Sci 38:567–572

Ford RE, Furmidge CGL (1967) Impact and spreading of spray drops on foliar surfaces. In: Wetting SCI (ed) Monograph no 25. Society of Chemical Industry, London, pp 417–432

Göhlich H (1992) Efficient application of herbicides. Proc 1st Int Weed Control Cong 1:150–158

Grayson BT, Webb JD, Pack SE, Edwards E (1991) Development and assessment of a mathematical model to predict foliar spray deposition under laboratory track spraying conditions. Pestic Sci 33:281–304

Grayson BT, Pack SE, Edwards E, Webb JD (1993) Assessment of a mathematical model to predict spray deposition under laboratory track spraying conditions. II. Examination with further plant species and diluted formulations. Pestic Sci 37:133–140

Hall FR (1990) Controlled delivery and foliar spraying. In: Wilkins RM (ed) Controlled delivery of crop-protection agents. Taylor and Francis, London, pp 3–21

Hall FR, Chapple AC, Downer RA, Kirchner, LM, Thacker JRM (1993) Pesticide application as affected by spray modifiers. Pestic Sci 38:123–134

Hartley GS (1967) Behaviour of water on surfaces at extreme limits of contact angles. In: Wetting SCI (ed) Monograph no 25. Society of Chemical Industry, London, pp 433–442

Hartley GS, Brunskill RT (1958) Reflection of water drops from surfaces. In: Danielli JF, Parkhurst KGA, Riddiford AC (eds) Surface phenomena in chemistry and biology. Pergamon, Oxford, pp 214–223

Holloway PJ (1993) Adjuvants for agrochemicals: why do we need them? Medelingen van de Faculteit Landbouwwetenschappen Rijksuniversiteit Gent 58/2a:125–140

Holloway PJ, Hayes AL, Western NM (1993) Influence of adjuvants on the retention and spreading of spray droplets on foliage (in preparation)

Lake JR (1977) The effect of drop size and velocity on the performance of agricultural sprays. Pestic Sci 8:515–520

Lefebvre AH (1993) Droplet production. In: Hislop EC, Matthews GA (eds) Application technology for crop protection. CAB International, Oxford, pp 35–54

Miller PCH (1992) Herbicide application, vol 1. Proceedings of the 1st international weed control congress, pp 159–164

Reichard DL (1988) Drop formation and impaction on the plant. Weed Technol 2:82–87

Spillman JJ (1984) Spray impaction, retention and adhesion: an introduction to basic characteristics. Pestic Sci 15: 97–106

Stevens PJG, Kimberley MO, Murphy DS, Policello GA (1993) Adhesion of spray droplets to foliage: the role of dynamic surface tension and advantages of organosilicone surfactants. Pestic Sci 38:237–246

Tadros TF (1987) Interactions at interfaces and effects on transfer and performance. Aspects Appl Biol 14:1–22

van Toor RF, Hayes AL, Cooke, BK, Holloway PJ (1993) Relationships between the herbicidal activity and foliar uptake of surfactant-containing solutions of glyphosate applied to foliage of oats and field bean. Crop Protect (in press)

Wirth W, Storp S, Jacobsen W (1991) Mechanisms controlling leaf retention of agricultural spray solutions. Pestic Sci 33:411–420

Young BW (1980) The use of high speed cine photography in pesticide spray application research. Proceedings of the 5th International Conference on Mechanization of Field Experiments, Wageningen, pp 116–123

Young BW (1986) The need for a greater understanding in the application of pesticides. Outlook Agric 15:80–87

5 Factors Affecting Penetration of Compounds Through the Cuticles of Insects and Mites

W. T. Lankford

5.1 Introduction

A casual perusal of promotional literature for insecticides reveals that the efficacy characteristics attributable to formulation composition are rarely praised. Indeed, the relative merits of the product are often credited solely to the insect control agent. Although product differentiation by active ingredients has undoubtedly been commercially successful, even where compounds have many similar competitors, this approach may have blunted commercial pressure to optimise formulations for biological activity. Such a suspicion is aroused by concern in the industry that the considerable increases in laboratory potency of new insecticides are not reflected in comparative use rates of the formulated product in the field. Evidence for this disparity was first discussed by Hartley and Graham-Bryce (1980). Averaged over four pest insect species, the relative toxicity of DDT compared with deltamethrin in the laboratory was 1600 to 1. However, in field use the recommended ratio of active ingredient was 50 to 1. The endeavour to reconcile this disparity highlighted the extremely low utilisation of insecticides at the site of action, often 0.1%, after placement in the target environment. Graham-Bryce (1984) also suggested that the processes which attenuate the applied dose have proportionately greater effects as rates of application are reduced.

In the early 1990s there has been increasing interest in optimisation of biological activity due to the need to reduce application costs of newer, more expensive active ingredients and also to promote the development of existing compounds as fewer compounds achieve registration (Holden 1992). Weaknesses in the dose transfer efficiency of crop protection applications are not difficult to find; for example, Hall et al. (1993) highlights them in detail. In broad subject areas efficiency gains that might result in significant dose reductions of active ingredient in an insect control programme may arise from three areas of research: understanding the biology of the pest and its natural control agents to optimise spray timing, developing application techniques to maximise the amount of formulation contaminating the habitat of the pest, and addressing the factors which affect penetration of formulated compounds through the cuticles of insects and mites.

The subject of insecticide uptake is usually approached from the viewpoint of the organic chemist, evaluating structure-activity relationships to provide leads for optimising insecticide conformation. Another common approach is the consideration of uptake in reviews of overall application efficiency. Both of these viewpoints help to build a holistic picture of the events leading to satisfactory insect control in which formulation composition plays a key part. Nevertheless formulation chemists must be confident that compounds for development have been optimised for pharmacokinetic and pharmacodynamic activity. In addition, the application technique must place the product in the same space and at the same time as the pest insect. Work can then concentrate on increasing uptake of the control agent, aware that improvements are likely to increase the biological effect. With contact insecticides in mind, this contribution reviews the penetration of compounds into arthropods and proposes opportunities for the optimisation of formulated insecticide products.

5.2 The Target Factor

5.2.1 Structure of Insect Cuticles

The interface between insects and their environment, the cuticle, is a multilayered extracellular composite structure. Borne from epidermal cells, cuticle extends over the body as part of the integument, internally facing apodemes form supporting structures for some muscles and internal organs. It covers all external appendages and, invaginated, it forms the lining of the oesophagus, foregut, hindgut, rectum, tracheae and tracheoles in addition to various glands of epidermal origin that open onto the surface (Filshie 1982).

The outermost zone, the epicuticle, is a chitin-free region. Usually 1 μm thick, it consists of a cement layer (in some species), then epicuticular wax overlying a layer of cuticulin.

Cuticle lipids encompass a wide range of compounds and wax secretions vary between insect classes (Richards 1978). In whiteflies triacylglycerols are the most abundant lipid class but in most other arthropods these are rarely found whilst hydrocarbons are prevalent (Byrne and Hadley 1988). Cuticle hydrocarbons may be branched, unbranched, saturated or unsaturated and are sometimes markedly different even in the same genus (Jackson 1972). Small quantities of fatty acids and alkyl esters are of frequent occurrence and sterols, alcohols, aldehydes and phospholipids have also been recorded in extracts of "free" cuticle lipids. Lipids are also "bound" in the cuticle structure and are not readily removed from lipoprotein layers by solvent extraction (Richards 1978). Using light microscopy and staining techniques Wigglesworth (1990) has observed the incidence of lipids and polyphenols in cuticles of a number of insect species. In regions where tubular filaments carrying polyphenols (sclerotin precursors) discharge alongside other tubular filaments secreting lipids, cuticulin is formed by the association of sclerotin with lipid. Three sites for cuticulin are known. The thin water-proofing layer on the outer epicuticle, in the substance of the amber-coloured inner epicuticle and in the hard amber-coloured laminated exocuticle.

Beneath the epicuticle are regions of the *N*-acetyl glucosamine polymer chitin (usually 25%–40% cuticle dry weight; Richards 1978), proteins (usually 50% cuticle dry weight; Richards 1978) and tanning agents which when laid down are collectively termed the procuticle and constitute about 90% of cuticle thickness. epidermal cells and is more hydrophilic than the other layers (Hassal 1982). Between the exo- and endocuticle lies the mesocuticle, which may be a further physical and chemical modification of endocuticle (Filshie 1982).

A degree of functional affinity between the structure of insect and plant cuticles has been noted by Hassal (1982). This is not surprising since the evolutionary pressures for integument function are the prevailing environmental conditions, common to all organisms occupying the same habitat. The important similarity for researchers of chemical uptake through insect cuticle is that the surfaces of both plants and insects present a hydrophobic target to formulations. Obviously the significant body of work on plant cuticles may therefore provide some useful references. We also should not ignore the prevalence of hydrophobic cuticle in other organisms, and the literature on topical applications of pharmaceuticals and cosmetics to mammal skin might be a further source of valuable information.

5.2.2 Movement of Naturally Occurring Substances Through Insect Cuticles

The structure of the cuticle provides a barrier to loss of water from terrestrial arthropods. Its effectiveness is beyond doubt, as their small volume to surface area ratio renders these animals highly susceptible to desiccation. Water may not move into or out of the cuticle, but the processes of cuticle renewal, moulting and sclerotisation, reveal that some compounds can migrate through the structure. Most insects possess transverse canal networks which directly connect the epidermis with various levels in the cuticle including the surface. In the chitinous part of the cuticle, these are pore canals, formed when cytoplasmic projections originating in the epidermal cells and traversing the chitinous procuticle to various levels are withdrawn or become embedded. The shape of pore canals is often determined by the orientation of surrounding cuticle microfibrils. In the epicuticle, a transverse canal or filament system is generally present, and this connects pore canals with the cuticulin layer (Filshie 1982). Epicuticular filaments are always present in cuticles which secrete wax and are usually seen in micrographs as simple cylindrical rods or tubules. When pore canals are not present, for example, in larvae of *Lucilia cuprinia* (Filshie 1982), wax precursors may pass from epidermal cells directly through the cuticle matrix and beyond via Chitin microfibrils aggregate into microfibres that are arranged predominantly in a highly organised helicoidal array (Reynolds 1987). The density and orientation of microfibres as well as chemical modifications identify distinct layers which are clearly visible by microscopy. Amino acids are found in association with chitin residues, indicating strong protein-chitin links. Tanning agents, such as tyrosine stabilise the cuticle and increase its hardness by causing sclerotisation. This cross-linking process occurs in the exocuticle and occasionally extends into underlying chitin layers resulting in increased hydrophobic properties. Endocuticle adjoins the epicuticular filaments.

The epidermal tissues of the cuticle are separated from the haemolymph by a basement membrane. The basal laminae is the primary barrier to the movement of macromolecules between tissues and the haemolymph (Reddy and Locke 1990). Movement of macromolecules across the laminae is determined by their charge and porosity. Examining stained tissues of *Calpodes ethlius* larvae that had been injected with colloidal gold of known size distribution, Reddy and Locke (1990) related uptake to the permeability of tissue basal laminae to haemolymph proteins. The epidermis, along with the fat body and pericardial cells, were permeable to particles 15 nm, whilst tissues not involved in the synthesis of haemolymph proteins were not permeable to the smallest (6 nm) particle. Once particles had crossed the laminae into tissues, their widespread distribution suggested no further filtration steps, although the authors were unsure if this was true for in vivo protein filtration. Haemolymph bathes the epidermal tissue and is the primary means by which materials are transported from sites of absorption or synthesis to the storage organs and, subsequently, to sites of utilisation (Downer and Matthews 1976). Compounds in the haemolymph may be involved in transport between tissues. For example, lipids may absorb growth hormones and, in the studies of Reddy and Locke (1990), artificial haemolymph had to be used because the gold colloids bound strongly to haemolymph proteins. The lipid profile in haemolymph varies with physiological state (Downer and Matthews 1976), but may reach concentrations of 5.5% total lipid, primarily comprising of acylglycerols with some triacylglycerols and free fatty acids.

5.3 The Toxicant Factor

5.3.1 Cuticular Penetration by Insecticidal Compounds

Given the complexity of insect cuticle, exhibiting hydrophobic layers overlaying non-living and living hydrophilic regions, it might be reasonable to expect it to be a formidable barrier to topically applied insecticides. In fact, the data on commercially successful compounds applied in organic solvents tends to suggest that it is an easily overcome barrier. Comparisons of ratios between the dermal and intravenous LD_{50} values of several insecticides, including examples of organochlorines and organophosphates, as well as pyrethrins and rotenone, for insects and mammals by Hassal (1982), show that dose effects in insects are rarely reduced by the cuticle barrier. In mammals, the LD_{50} dose is usually far higher for dermal applications, possibly due to a greater barrier effect but it is equally likely to reflect a higher metabolic capacity in mammalian integument, resulting in significant detoxification at this site. Soderlund (1980) interpreted the usually small differences between toxicity of topical and injected insecticides as evidence that penetration is not an important rate-limiting step in the intoxication process. Reduced uptake, for example, in housefly strains with the resistance gene *pen*, has little effect on toxicity, unless another resistance mechanism is present.

A wide range of physicochemical properties confer access of compounds into arthropod bodies. The polarity of commercially successful chemical control agents, indicated by their octanol-water partition coefficient (P), ranges from log P –1.0 to log P 7.0 (Briggs et al. 1974). A spectrum which at its low end is characterised by compounds that show good systemic properties in plants and which may be effective against foliar pests when soil-applied. Progressing through the mid-range of log P there is some systemic action in plants, whilst non-polar compounds (log $P > 4$) are contact-acting only. In synthetic pyrethroids good knockdown and good kill arise from common structural features which allow a good fit at active site(s) (Briggs et al. 1974). Whether or not the compound expresses good knockdown is also influenced by polarity, with more polar compounds showing more rapid onset of symptoms. This is a relationship commonly interpreted as a consequence of faster penetration by the more polar pyrethroids. This is not surprising. Although the first barrier is a lipid layer and epicuticular lipid filaments offer a hydrophobic pathway to the epidermal cells via pore canals, at the base of the cuticle, non-polar compounds would not partition favourably into aqueous media, particularly that of the haemolymph. The polarity of a compound also has ramifications for absorption into insect tissues, including non-target organs. For a range of aliphatic alcohols, penetration from Ringer bathing solution into American cockroach nerve cords was negatively correlated with polarity (Eldefrawi and O'Brien 1967), although compounds with log $P > 3$ did not display greater penetration.

Although not immediately important for the formulation of contact insecticides, knowledge of penetration by the active ingredient alone may be useful if situations are anticipated where contact occurs in the absence of carrier solvent, for example, compounds with a long residual life or significant activity in the vapour phase.

5.3.2 Possible Uptake Pathways

There is longstanding controversy about the pathways by which compounds reach their site of action through arthropod cuticles. The majority of workers consider that insecticide uptake is similar to the mammalian model of pharmacokinetics, where compounds diffuse through the integument and are distributed in the animal by the haemolymph. An alternative view has been proposed by Gerolt (1969) working principally with dieldrin. He considers that compounds diffuse laterally through the cuticle and arrive at the site of action via the trachea. The supporting arguments for both sides have been reviewed by Welling (1977) and Welling and Paterson (1985); they found no conclusive evidence for either hypothesis. Although the lateral transport theory remains plausible, good evidence exists for distribution of insecticides via the haemolymph. The pyrethroid compounds NRDC 157 and NRDC 163 (Soderlund et al. 1983) and cypermethrin (Greenwood et al. 1990; Lankford and Dawson 1993) are detected in the haemolymph soon after topical application. Tracheal uptake is perhaps the least credible origin of these initial haemolymph titres. For whilst the insect circulatory system is of an open nature, bathing internal organs, supplying nutrients and removing metabolic wastes, it does not play a major role in respiratory processes (Gilbert and Chino 1974). Cuticular absorption is likely to be important, as revealed by Bruck and Stockem (1976). Their histochemical analyses show that the organophosphorous insecticide, triazophos, is absorbed into the treated pronotum of cockroaches and is highly localised in pore canals in the cuticle.

5.4 The Influence of Formulation

5.4.1 The Effect of Insecticide Carriers

The published evidence of insecticide penetration has led some authors discussing this subject (Soderlund 1980; Hassal 1982) to suggest that most commercial insecticide compounds penetrate insect cuticle with relative ease. In fact, this may only be partially true and it is a complacent assumption for formulation chemists to adopt. In the majority of original studies (for example Hadaway et al. 1976; Holden 1979; Soderlund 1980; Soderlund et al. 1983; Greenwood et al. 1990), the insecticide was formulated in an organic solvent. Not only is this a paradox for industry researchers when ideal formulations are considered to be solvent-free (Mulqueen 1990) but, moreover, other than water it is doubtful whether any topical application fluid is a benign carrier for insecticides. It is illuminating that phosphoric acid applied in acetone has been found to penetrate quickly into *Periplaneta americana*, while phosphoric acid dissolved in water was not absorbed until all of the water had evaporated (Olsen and O'Brien 1963).

Lewis (1980) warns that acetone solubilises wax into the applied droplet, and although it quickly evaporates, the insecticide is dissolved in as well as on the redeposited wax, effectively by-passing the barrier effect of the epicuticular wax layer. Three processes characteristic of the formulating solvent influence penetration (Lewis 1980): surface migration, interaction with epicuticular lipids and concentration of the solvent. Codiffusion through the cuticle with insecticide is a further potential effect of the formulating solvent. However, using conventional "wash-off" techniques, the flux of two compounds through the cuticle surface is seemingly determined for each compound by its own physical characteristics. Lewis (1963) found that dieldrin had a faster penetration rate than its solvent hexadecane, whilst Lankford and Dawson (1993) revealed that toluene had a faster penetration rate than its solute, cypermethrin.

The properties of the carrier solvent can strongly effect the rate of penetration. DDT penetrates most rapidly into tsetse flies when its solvent is of low viscosity (1.14 mN/m^2, 25°C) and has a moderate (130°–270°C, 760 mm Hg) boiling point range (Hadaway et al. 1976). Solvents with higher viscosity spread more slowly, contaminating a smaller area of the cuticle with the marker dye. Internal DDT recovery was lowest at each time point for carriers with low volatility (boiling point 300°C, 760 mm Hg), and this is correlated with low toxicity at 24 h. Loss of volatile compounds is reduced by oils of low volatility, but so is toxicity. In the same investigation, Hadaway et al. (1976) recorded large differences between the penetration of DDT in different species over a 4 h time course. After 4 h the dose which had penetrated ranged from 9% in *Blattella germanica* to 80% in *Stomoxys calcitans*.

For aqueous based applications, penetration of insecticides is probably inversely related to polarity (Treherne 1957). This suggests that the hydrophobic wax layers of the epicuticle are the significant barrier. For dimethoate, DDT, dieldrin, paraoxon and phosphoric acid and dipotassium hydrogen phosphate applied in organic solvent, speed of penetration is apparently related to polarity (Olsen and O'Brien 1963). However, Olsen and O'Brien (1963) believed that for their applications in small volumes of volatile solvent, epicuticular wax was relatively unimportant in controlling penetration of solutes. In a later review of penetration studies, Welling and Paterson (1985) concluded that for a polar solvent, partitioning of insecticide into the epicuticular lipids can be the limiting process.

Quantitative experiments analysing the effects of adjuvants on uptake are rare, but one study of diflubenzuron (DFB) applied to *Spodoptera littoralis* (El Saidy et al. 1988) indicates the potential in this area. All four of the commercial adjuvants tested at a concentration of 0.3% significantly increased absorption of topically applied DFB. In the two adjuvant combinations of paraffinic oil and surfactant, the foliar LC_{50} was halved. Although there was no relationship between effects on topical absorption and residual toxicity, alkoxylated alcohols and volatile silicones doubled uptake but increased the LC_{50}. In a topically applied microemulsion formulation, initial uptake of cypermethrin is likely to be as part of the intact system (Lankford and Dawson 1993). Prolonging the stability of the microemulsion, post-application, may increase uptake of the insecticide. Increasing the rate of uptake does not always result in greater efficacy. In the case of cypermethrin in susceptible strains of insects, the active isomers are relatively stable to metabolism (Holden 1979), and excretion rates are low, leading to accumulation of the toxicant even at slow rates of penetration.

5.4.2 Detecting Opportunities for Formulation Enhancement

As a consequence of difficulties in achieving registration of new compounds, opportunities now exist for improving the efficacy of existing compounds (Holden 1992). How can the best of these opportunities be identified? Information on the pharmacokinetic profiles for compounds applied in a commercially relevant formulation are rare in the literature. There are excellent reviews of the appropriate methodology (Lewis 1980; Welling and Paterson 1985), including interpretation of efficiency in absorption and partition processes. In vivo systems are preferred by many workers because insect activity influences contamination, for example, the movement of surface films across the cuticle (Lewis 1963). On the other hand, for formulation chemists accustomed to minimal variation in data, isolated cuticle systems have the advantage of a potentially high degree of standardisation. Probably the best simulation of in vivo uptake occurs when retaining the epidermis of the integument. The outer plasma membrane of the epidermal cells controls movement of chemicals into the cuticle, and the basement membrane limits transport between the haemolymph and epidermal cells (Richards 1978). Metabolism in these cells may also effect absorption rates of insecticides (Lewis 1980).

Suitable labelled and non-labelled techniques are employed in the evaluation of plant and animal metabolism and in the evaluation of environmental fate of compounds for registration purposes. Formulation scientists could cooperate with these research groups for advice and equipment time. Furthermore, they may already have experience in deriving models for analysing the behaviour of compounds. Mathematical models show the quantitative changes in penetration which are required for a significant effect at the target site (Greenwood et al. 1990). Further formulations could be assessed by measuring time-related changes in a limited number of compartments.

Whilst the published studies of insecticide penetration suggest that most neurotoxic compounds show high levels of absorption through the cuticle when applied in organic solvent, this is not true for all chemical control agents. The 24 h penetration of DFB applied in acetone to *Spodoptera littoralis* was 4.9% (El Saidy et al. 1988). In addition, poor cuticle penetration rates have been quantified for other acylurea compounds. The high recoveries of acetone-dissolved flufenoxuron and teflubenzuron in solvent washes of topically treated *S. littoralis* and *S. exempta* (Fisk and Wright 1992) also infer poor penetration into chitin containing regions and epidermal cells, which, according to Reynolds (1987), are the likely sites of action for acylurea compounds. For these acylureas, mechanistic studies of uptake by insects suggest good opportunities for biological optimisation by formulation development.

5.4.3 Optimising the Biological Effect of Formulations

While this paper concentrates on the uptake of compounds at the time of application and during droplet drying, formulators must be aware that improvements in activity can also arise from firstly optimising retention of the formulation in the target area. This is particularly important as the behaviour of insects ensures that they are discrete targets in space and time. In controlled droplet applications, direct contact of spray droplets with sessile insects is probably uncommon. Munthali and Scopes (1982) showed that placing 200 droplets in a 1 cm^2 area results in 10% direct hits on eggs of the red spider mite *Tetranychus urticae*. Hall (1988) supports this observation, stating that in his laboratory direct hits are frequently 10% irrespective of "good" coverage of the target leaf. Direct contact of droplets containing concentrations of toxicant in excess of the LD_{100} is a waste (Graham-Bryce 1984). Formulators should also consider that, for residual treatments, the surface of the target plays a key role in treatment efficacy. Dose transfer from plant surfaces has been reviewed by Ford and Salt (1987), whilst Chadwick (1985) has reviewed formulation interactions with surfaces in public hygiene applications.

Extrapolating from studies using organic solvent applications of insecticides provides a good background for speculation on the ideal properties for effective aqueous formulations. Low surface tension and viscosity increases cuticle contamination. A solubilising capacity for aliphatic waxes for example, C_{25}–C_{43} paraffins, long-chain alcohols, fatty acids and alkyl esters may improve the uptake of less lipophilic substances by negating the partition step from the aqueous droplet into wax. On the potentially negative side, very low volatility formulants with a capacity to solubilise the active ingredient may render it unavailable for uptake. In cases where contamination is by body contact with deposits on the plant cuticle (Ford and Salt 1987), this phenomenon confers low volatility oil formulations of insecticides with relatively high residual toxicity.

Agrochemical formulators find that few structural groups from the vast range of surfactant types available are useful, affordable and adequately supported for registration purposes. For series of compounds that meet requirements it is, however, worthwhile screening for effects on efficacy. Surfactant manufacturers often use hydrophilic-lipophilic balance (HLB) values to distinguish blends in a chemical series. Whilst researchers investigating adjuvant activity have made some good progress on identifying some effects of surfactant HLB on herbicide uptake, much less is known about HLB effects on the activity of insecticides. In microemulsion formulations of pyrethroids, researchers at NCH Europe have found that low HLB cosurfactants show lower, but never significantly different relative activity against cockroaches compared with flies (Table 1). They are the least potent combination against glasshouse whitefly (Table 2) and have poor initial activity when sprayed on to caterpillars of *Pieris brassicae* (Table 3). Cypermethrin and deltamethrin formulations in the cosurfactant series were more toxic to glasshouse whitefly scales than EC standards but less effective against caterpillars of *P. brassicae*. The prospective formulator might be disappointed that there are no clear-cut HLB-activity relationships for this cosurfactant series. However, cosurfactant choice resulting from other criteria would not compromise activity, and further biological development towards targeting small sessile pests has potential.

The screening of commonly used formulations and their components may reveal surfactants or solvents with useful biological effects. Petroleum oils and fatty acids both show contact activity against small arthropod pests. Their mode of action may involve suffocation and/or desiccation. Wigglesworth (1990) points out that solvents and detergents often promote water-loss when applied to the insect cuticle. Combinations of formulants which form amphiphilic association colloids are known to cause effects in cuticles which relate to their structure. Intact microemulsions may solvate the ordered lipid structure in human stratum corneum, forming a liquid and consequently causing functional disruption, with associated irritation due to facilitated transdermal transport of the xenobiotics (Friberg 1990). Water rapidly penetrates and evaporates from microemulsions applied to insects (Lankford and Dawson 1993) and human skin (Friberg 1990). Therefore, the extent of the disruption is related principally to the structure of the residue after formulant evaporation. Oil solutions of surfactants show increased irritancy compared to liquid crystal residues (Friberg 1990). If similar interactions occur in the insect integument, formulations containing high boiling point solvents might best enhance the disruption of the waterproofing function of cuticular lipids. Water loss would be further aggravated if there is water absorption through the cuticle into persistent deposits of microemulsion formulants. In a mammalian test system, water-free microemulsion bases absorbed water forming supersaturated microemulsions (Kemken et al. 1991). The resulting thermodynamic activity and decreasing capacity for the lipophilic β-blocker promoted uptake of the drug.

Lankford and Dawson (1993) found that water in an oil-in-water microemulsion was rapidly absorbed into and through cockroach cuticles after in vivo application, an unexpected result considering the intrinsic waterproof properties of this cuticle. It was suggested that water uptake was a consequence of damage to wax layers caused by lipid solvation into the disperse phase of the formulation. Perhaps this indicates how a small volume of solvent in a water-based product might disrupt the epicuticular wax barrier of the insect cuticle and thus facilitate the ingress of polar compounds into less hydrophobic regions of the cuticle. This phenomenon has also been observed for some topical formulations of pharmaceuticals. Osborne et al. (1991) found that on human skin, uptake of hydrophilic drugs depends, firstly, on the mobility of water in the microemulsion vehicle and, secondly on sufficient percutaneous transport of water across the skin barrier. Although an unconventional approach, one potential ramification of the data of Osborne et al. (1991) is that hydrophilic compounds might be efficiently formulated in oil-in-water systems.

5.4.4 Routine Toxicity Assays of Contact Insecticide Formulations

Formulation scientists, for example, Franklin and Hartmann (1990), often employ empirical approaches in product development and are well experienced in rapidly producing variations in their formulations. To find efficient partners for these skills, biologists need to be involved. They must communicate the properties of the formulation which they believe to be important for efficacy, discuss how these features may be addressed, and then provide their data quickly so that efficacy is on the agenda when stability properties are reviewed.

The requirement for haste in assessing biological activity and cost concerns, where large numbers of candidate formulations may be involved, dictate that toxicity assays are the method of assessment. Bioassays do not solely reflect penetration of compounds but demonstrate toxicity as a consequence of all the modifying effects in the pharmacokinetic process. This approach therefore is a compromise. Worthwhile results are more likely to be obtained where knowledge of a compound's (or related structures) pharmacokinetics is available. For example, is there capacity in the target for effective distribution of increased doses? Soderlund (1980) found that the distribution of NRDC 157 enantiomer II is proportional to the applied dose in haemolymph, nerve cord and fat

Table 1. Influence of cosurfactant HLB on the toxicity of cypermethrin microemulsions to public hygiene pests (modified from NCH Europe contract report LC/89/0043, unpublished), assessment of the efficacy of cypermethrin and bioresmethrin ready-to-use formulations against *Musca domestica*

HLB value of cosurfactant	KD_{50} (± 95% CI) ng/fly *Musca domestica*	KD_{95} (± 95% CI) ng/fly *Musca domestica*	KD_{50} (± 95% CI) ng/cockroach *Blattella germanica*
1	1.7 (1.5–1.8)	5.2 (4.4–6.4)	150 (128–172)
1 (+ 0.2% WT10)	1.3 (1.1–1.4)	6.4 (5.1–8.8)	
2	–		
3	–		154 (131–177)
5	2.0 (1.9–2.2)	6.1 (5.2–7.4	123 (110–136)
7	–		132 (113–151)
12	–		144 (124–164)

CI, confidence interval; WT10, Tegoplant WT10, Goldschmidt. Germany; KD_{50}, dose of insecticide required per insect to knockdown 50% of the insect population.

body up to a dose of 17 μg/g, suggesting that even for the most lipophilic pyrethroids, internal bioavailability is not compromised.

In Tables 1–3, examples of toxicity assays carried out on various microemulsions of pyrethroids reveal how the same formulation variable may be more or less important in effecting toxicity depending upon the toxicant, test type and insect species. For compounds such as deltamethrin which are poorly metabolised (Soderlund 1980), prolonged uptake does not reduce the lethal effect. The deltamethrin formulation with the poorest knockdown at 24 h shows the highest toxicity after 3 days (Table 3). Assessments should also evaluate the slope of the biological response because the best formulation at LC_{50} may not be ranked the same at LC_{90} (Table 1). Assay design should include assessments of time to the onset of biological effect as well as the dose related biological effect and its persistence. Sub-lethal doses are potentially disruptive to insect populations (Haynes 1988), and it is likely that the economic benefits from insecticide treatment may not always be directly related to mortality. It therefore makes sense to include observations of insect

Table 2. Influence of cosurfactant HLB on the lethal concentration of pyrethroid formulations to first instar larvae (scales) of *Trialeurodes vaporarorium* (modified from NCH Europe report C/88/0353/4–6, unpublished), tests of novel formulations of pyrethroid insecticides against some phytophagous pests

HLB value of cosurfactant	Deltamethrin		Cypermethrin	
	LC_{50} ppm AI	LC_{90} ppm AI	LC_{50} ppm AI	LC_{90} ppm AI
EC standard	2.4	100	13.5	300
1	1.9	25	–	
2	0.7	13	15.5	139
3	1.9	14	13.5	131
7	1.5	14	–	
3 (+ wetter)	–		9.3	156

First instar larvae on bean *Phaseolus vulgaris* leaves (three replicates) were dip-treated (nine concentrations). AI, active ingredient. Standards: deltamethrin EC: "Decis", Hoechst. UK; cypermethrin EC: Ambush C, ICI. UK; wetter: 2% Tegoplant WT10, Goldschmidt, Germany; LC_{50}, concentration of insecticide in treatment solution required to kill 50% of the insect population.

behaviour in laboratory bioassays of formulation and this has been demonstrated for cyhexatin and some pyrethroids by Hall (1988). In several comparative tests of insecticides against *T. urticae*, using assessments of activity include behaviour, egg laying, feeding damage and mortality, Hall (1988) observed examples of enhancement of interruptive effects on mite behaviour that were related to formulation but not associated with significant improvements in lethality.

Information on the distribution and retention of formulations must be gathered to fully understand the data gained from screening. Together with information on crop growth and pest biology this has recently been assembled into a model (Day and Collins 1992) with considerable promise for researchers trying to optimise insecticide treatments.

Table 3. Influence of microemulsion cosurfactant HLB on the toxicity of direct sprays to *Pieris brassicae* (modified from NCH Europe report C/88/0353/4–6, unpublished), tests of novel formulations of pyrethroid insecticides against some phytophagous pests

HLB value of cosurfactant	LC_{50} (± 95% CI) (ppm AI) after 24 h	LC_{50} (± 95% CI) (ppm AI) after 3 days
Deltamethrin		
"Decis" EC	0.6 (0.4–0.9)	0.5 (0.4–0.7)
1	4.3 (3.0–6.2)	0.7 (0.4–0.9)
2	2.2 (1.6–3.1)	1.7 (1.2–2.5)
3	1.4 (1.0–2.0)	1.2 (0.8–1.7)
7	1.9 (1.3–2.7)	1.8 (1.3–2.6)
Cypermethrin		
"Ambush" C EC	1.3 (0.9–1.8)	1.6 (1.1–2.4)
2	7.3 (5.2–10.4)	2.8 (2.0–4.1)
3	3.5 (2.5–5.0)	1.5 (1.0–2.3)
3 (+0.2% WT10)	9.1 (6.5–12.9)	4.0 (2.8–5.8)

Second instar larvae (three replicates of ten) were direct-sprayed (Potter tower, five doses), then transferred to untreated cabbage leaves for observation. AI, active ingredient; LC_{50}, concentration of insecticide in spray solution required to kill 50% of the insect population.

5.5 Practical Considerations for Commercial Products

The fundamental role of the pesticide formulation chemist is always to develop formulations which are both physically and chemically stable and facilitate the easy application of the active ingredient. Good quality and dependable products are vital. Optimum biological efficacy should be a priority close behind these aims, but to move it higher on the list of formulation priorities requires determination and discussion of the potential impact on other aspects of the product. Marketing specialists need assurance and definition of any unique features, whilst business managers must be convinced that low costs in raw materials and production will come from reduced application rates. Unfortunately, the high costs of individual product registrations ill effectively limit the number of formulations for any one compound, market or application technique. If this is the case the development of better adjuvants for tank-mixing with insecticide products may be a more cost-effective solution. Whether researchers select formulation modification or adjuvants for tank-mixing as their favoured approach to optimising insecticides, work should begin early in the development life of candidate compounds. One reason is that if biological optimisation is left until the late stage of field trials, the effects of temporal and spatial placement of the product may blur conclusions about the formulations intrinsic activity. Many opportunities for identifying commercially useful compounds may have been missed in the past (Graham-Bryce 1979), especially during the initial screening of relative insecticide potency using simple formulations. Where the physicochemical properties of compounds do not favour penetration into insect cuticle, scope exists for formulation effects to optimise uptake, and there is good evidence that any missed opportunities could be minimised in the future.

Acknowledgements. P. Chapman at C.S.L, Slough, UK carried out the housefly tests. L. Wardlow and J. Cross at ADAS, Wye, UK carried out toxicity tests against whitefly and cabbage white caterpillars. Thanks are also due to L.G Davies and D. Stock for reading the draft manuscript and offering constructive advice.

References

Briggs GG, Elliot M, Farnham AW, Janes NF (1974) Structural aspects of the knockdown of pyrethroids. Pestic Sci 5:643–649

Bruck E, Stockem W (1976) Histochemical localisation of triazophos in the cuticle of cockroaches. Z Angew Entomol 81:136–140

Byrne DN, Hadley NF (1988) Particulate surface waxes of whiteflies: morphology, composition and waxing behaviour. Physiol Entomol 13:267–276

Chadwick PR (1985) Surfaces and other factors modifying the effectiveness of pyrethroids against insects in public health. Pestic Sci 16:383–391

Day RK, Collins MD (1992) Simulation modelling to assess the potential value of formulation development of lambda cyhalothrin. Pestic Sci 35:45–63

Downer RGH, Matthews JR (1976) Patterns of lipid distribution and utilisation in insects. Am Zool 16:733–745

Eldefrawi ME, O'Brien RD (1967) Permeability of the abdominal nerve cord of the american cockroach Periplaneta americana (L.) to aliphatic alcohols. J Insect Physiol 13:691–698
El Saidy MF, Auda M, Degheele D (1988) Comparison of four adjuvants on toxicity, absorption and residual activity of diflubenzuron on Spodoptera littoralis larvae. Proceedings of the Brighton Crop Protection Conference. Pests Dis 2:275–280.
Filshie BK (1982) Fine structure of the cuticle of insects and other arthropods. In: King RC, Akai H (eds) Insect ultrastructure. Plenum, New York, pp 281–312
Fisk T, Wright D (1992) Response of spodoptera exempta (Walk.) larvae to simulated field spray applications of acylurea insect growth regulators with observations on cuticular uptake of acylureas. Pestic Sci 35:321-330
Ford MG , Salt DW (1987) Behaviour of insecticide deposits and their transfer from plant to insect surfaces. In: Cottrell HJ (ed) Pesticides on plant surfaces. Critical reports on applied chemistry,vol 18, Wiley, London, pp 26–86
Franklin NC, Hartmann J (1990) Storage stability of pesticide formulations. Proceedings of the 7th International Congress of Pesticide Chemistry, vol 2, p 8
Friberg SE (1990) Micelles, microemulsions, liquid crystals, and the structure of stratum corneum lipids. J Soc Cosmet Chem 41:155–171
Gerolt P (1969) Mode of entry of contact insecticides. J Insect Physiol 15:563–580.
Gilbert LJ, Chino H (1974) Transport of lipids in insects. Journal of Lipid Research 15:439–455
Graham-Bryce IJ (1979) The role of the formulation in biological activity. In: Geissbuhler H (ed) Advances in pesticide science, vol 3. Pergamon, Oxford, p 717
Graham-Bryce IJ (1984) Optimization of physicochemical and biophysical properties of pesticides. In: Magee P, Kohn GK, Menn JJ (eds) Pesticide synthesis through rational approaches. American Chemical Society, Washington DC, pp 185–207 (ACS symposium series 255)
Greenwood R, Ford MG, Peace EA, Salt DW (1990) The kinetics of insecticide action. IV. The in vivo distribution of pyrethroid insecticides during insect poisoning. Pestic Sci 30:97–121
Hadaway AB, Barlow F, Flower LS (1976) Penetration of insecticides from solutions into tsetse flies and other insects. Centre Overseas Pest Contr Misc Rep 22:1–14
Hall FR (1988) Pesticide formulations and other parameters affecting dose transfer. In: Cross B, Scher HB (eds) Pesticide formulations – innovations

and developments. American Chemical Society, Washington DC, pp 260–278 (ACS symposium series 371)

Hall FR, Chapple AC, Downer RA, Kirchner LM, Thacker RM (1993) Pesticide application as affected by spray modifiers. Pestic Sci 38:123–133

Hartley GS, Graham-Bryce IJ (1980) Physical principles of pesticide behaviour. Academic, London

Hassal KA (1982) The chemistry of pesticides, 1st edn. Macmillan, London

Haynes KF (1988) Sublethal effects of neurotoxic insecticides on insect behaviour. Annu Rev Entomol 33:149–168

Holden JS (1979) Absorption and metabolism of permethrin and cypermethrin in the cockroach and the cotton-leafworm larvae. Pestic Sci 10:295–307

Holden WTC (1992) Future formulation trends — likely impact of regulatory and legislative pressure. Proceedings of Brighton Crop Protection Conference. Pests Dis 1:313–320

Jackson LL (1972) Cuticular lipids of insects. IV. Hydrocarbons of the cockroaches Periplaneta japonica and Periplaneta americana compared to other cockroach hydrocarbons. Comp Biochem Physiol 41:331–336

Kemken J, Ziegler A, Muller BW (1991) Investigations into the pharmacodynamic effects of dermally administered microemulsions containing β-blokkers. J Pharm Pharmacol 43:679–684

Lankford WT, Dawson HB (1993) Concurrent events in the uptake by insects of cypermethrin from a microemulsion formulation. Pestic Sci 39:85–94

Lewis CT (1963) Some applications of radioisotopes to the study of contamination of insects by insecticidal solutions. Proc Ser Int Atomic Energy Agency 74:135–146

Lewis CT (1980) The penetration of cuticle by insecticides. In: Miller TM (ed) Cuticle techniques in arthropods. Springer, Berlin Heidelberg New York, pp 367–400

Mulqueen PJ (1990) Surfactants for agrochemical formulations. In: Karsa DH (ed) Industrial applications of surfactants. II. Royal Society of Chemistry, Cambridge, pp 276–302

Munthali DC, Scopes NEA (1982) A technique for studying the biological efficiency of small droplets of pesticide solutions and a consideration of the implications. Pestic Sci 13:60–62

Olsen WP, O'Brien RD (1963) The relation between physical properties and penetration of solutes into the cockroach cuticle. J Insect Physiol 9:777–786

Osborne DW, Ward AJL, O'Neill KJ (1991) Microemulsions as topical drug delivery vehicles: in vitro transdermal studies of a model hydrophilic drug. J Pharm Pharmacol 43:451–454

Reddy JT, Locke M (1990) The size limited penetration of gold particles through insect basal laminae. J Insect Physiol 36:397–407
Reynolds SE (1987) The cuticle, growth and moulting in insects: the essential background to the action of acylurea insecticides. Pestic Sci 31:131–146
Richards GA (1978) The chemistry of insect cuticle. In: Rockstein M (ed) Biochemistry of insects. Academic, New York, pp 205–232
Soderlund DM (1980) Pharmacokinetic modifications of intrinsic insecticide activity. Insect neurobiology and pesticide action (Neurotox '79). Society of Chemical Industry, London 449–455
Soderlund DM, Hessney CW, Helmuth DW (1983) Pharmacokinetics of cis- and trans-substituted pyrethroids in the American cockroach. Pestic Biochem Physiol 20:161–168
Treherne JE (1957) The diffusion of non-electrolytes through the isolated cuticle of Schistocerca gregaria. J Insect Physiol 1:178–186
Welling W (1977) Dynamics aspects of insect-insecticide interactions. Annu Rev Entomol 22:53–78
Welling W, Paterson GD (1985) Toxicodynamics of Insecticides. In: Kerkut GA, Gilbert LI (eds) Comprehensive insect physiology, biochemistry and pharmacology, vol 12. Insect control, Pergamon, New York, pp 603–645
Wigglesworth VB (1990) The distribution, function and nature of "cuticulin" in the insect cuticle. J Insect Physiol 36:307–313

6 Adjuvants and the Activity of Fungicides

W. Steurbaut

6.1 Introduction

Pesticides are generally applied in the form of a formulated product which, after dilution with water, gives a spray solution. Although the active material must guarantee the biological activity, other constituents are needed to bring this pesticide in a form suitable for application. Such so-called formulation additives and/or adjuvants are very diverse, but solvents, surfactants, and oils are most frequently used.

Surface-active agents in particular are becoming increasingly important because they can produce a whole range of effects, influencing application, emulsion stability, wetting and sticking performance, rainfastness, penetration through the cuticle, etc. All of these effects have been investigated very thoroughly because they are of pivotal importance for efficient pesticide application. Herbicidal activity especially can be improved by these promoting properties, and this has resulted in an extensive number of publications on herbicide – surfactant interactions. In a bibliography of the world literature on adjuvants (Foy et al. 1992), more than 90% of the references referred to herbicide applications. Also insecticide – surfactant combinations are becoming of current interest because the activity of an insecticide can be optimized by improving contact due to better wetting and penetration of the insect cuticle, or to an intensified exposure during movement of insects on the treated surface. In addition, (eco)toxicological side effects may be influenced by the choice of appropriate adjuvants.

The influence of adjuvants on fungicide activity has less extensively studied than has that of herbicides or even insecticides. This is at first sight surprising because it is known that penetration and translocation of systemic compounds can be altered by adjuvants. It is likely therefore that in the future specific surfactant–fungicide research will become much more important.

6.2 Complexity of Fungicide-Adjuvant Interactions

Investigations into the possible advantageous effects of fungicide–adjuvant combinations have been of little interest to researchers. This is partly due to the fact that the first generation of fungicides had only contact activity. Thus, they acted only preventively by the formation of a uniform layer on the treated leaf surface. This was achieved only with a high volume application, sufficient wetting, and repeated applications.

The situation changed totally with the introduction of systemic fungicides, which can penetrate the plant and are transported and distributed via the apoplast. This eliminates the necessity of a complete protective surface treatment but still allows provision for both preventive and curative treatments. The penetration and redistribution of the active ingredient (AI) together with the advantageous influences of additives and adjuvants are becoming increasingly important. It is obvious that this can give rise to a whole range of interactions among the different

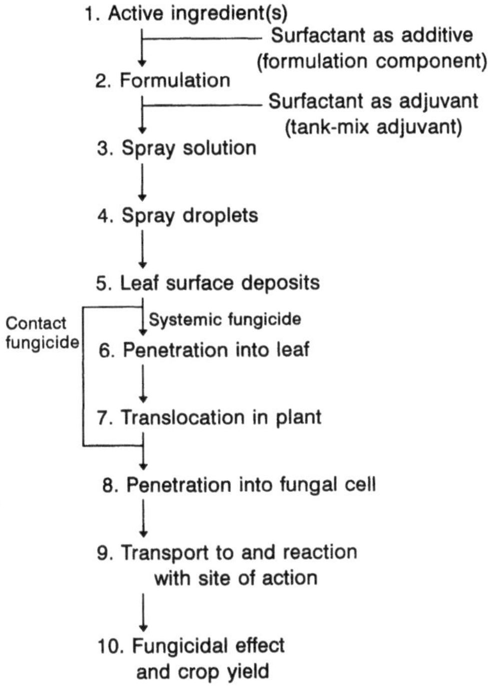

Fig. 1. Interrelationships of a fungicide–adjuvant combination on a plant-fungus system

parties involved – plant, fungus, fungicide, and surfactant. Such interrelationships can lead to far-reaching consequences, as illustrated in Fig. 1. Although a general pathway for pesticide action and the influences of surfactants can be drawn up (Steurbaut 1993; Fig. 2), there are some particular steps unique for fungicides.

The first steps (steps 1–5, Fig. 2) are valid for all foliar treatments with current pesticide formulations – emulsifiable concentrate (EC), wettable powder (WP), suspension concentrate (SC), and wettable granules (WG) – which form dispersions in the spray liquid. The influence of surfactants, as well as additives in the formulation as tank-mix adjuvants in the spray liquid, can be at several sites, including an effect

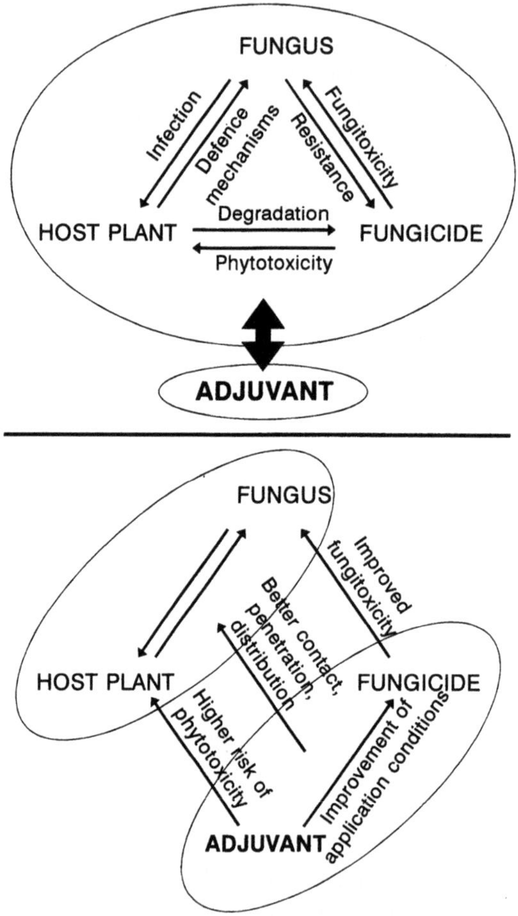

Fig. 2. Schematic presentation of fungicide activity

on the promotion of spray stability (emulsification, suspension stability), an influence on the wetting-spreading properties of the spray droplets on the leaf surface, improved sticking properties, and modification to the physical form of the drying deposit on the leaves. For systemic pesticides which must cross the physical barrier of the plant cuticle, other important factors are involved, such as optimum hydrophilic-lipophilic properties of AI and surfactant, influences of rewetting and the physical state of the deposit, in addition to the possibility of stomatal penetration (steps 6 and 7, Fig. 2).

For fungicides important extra steps are involved which can influence very dramatically their final activity (steps 8–10, Fig. 2). These are related to possible phytotoxic side effects, specific demands for fungal cell penetration, and a range of other interactions which can interfere with these steps, such as the behavior of infected plants to a treatment, the effect on resistance development of the fungus, the treatment type, and the location of the biochemical site of action in the fungal organism.

6.3 Influence of Surfactants on Penetration and Transport of the Active Ingredient

Penetration and transport processes (steps 6 and 7, Fig. 2) have been extensively studied for herbicides using isolated cuticles (Chamel et al. 1992; Chaumat et al. 1992; Bauer and Schönherr 1992). It is now generally agreed that these enhancement processes are the net result of a whole range of specific interactions between the AI, the surfactant, and the target plant species. For fungicides, however, fungus-related factors are also involved.

Despite all these studies the role of adjuvants in assisting the movement of systemic compounds within the plant is still not fully understood (Baker et al. 1992). It is still not possible to observe a clear contribution of surfactants to fungicide mobility, and there are still some conflicting results where, in particular situations, an interactive facilitation of translocation is possible. This occurs only when the surfactant penetrates the cuticle at the same or at a faster rate than the fungicide (Stock et al. 1992). In this way surfactant molecules can modify penetration routes and promote the transfer of fungicide molecules.

Experimentally, however, it is not easy to isolate the effects of penetration from transport and from the effects of improved spreading and coverage of the treated surface. Steurbaut et al. (1989) found a significant increase in the transcuticular penetration of fungicides in the presence of polyethoxylated nonylphenol (NP) surfactants but only when the mean molar ethylene oxide (E) was less that 10 units and at concentrations below 1% (Table 1). The effect on coverage was much

Table 1. Influence of nonylphenol surfactant E content and concentration on leaf coverage (percentage leaf area) and penetration of propiconazole (cuticular membrane penetration, percentage applied dose; from Steurbaut et al. 1989)

Treatment		Coverage[1]	Penetration[2]	Coverage× penetration
Control	–	24.1[a,b]	13.1[a,b]	3.2
NPE10	0.001%	18.8[a]	12.8[a]	2.9
	0.01%	40.0[c]	14.0[a,b]	5.6
	0.1%	98.7[e]	15.2[b]	15.1
	1.0%	99.0[e]	26.6[c]	26.5
	5.0%	99.0[e]	32.0[d]	32.0
NPE6	1%	99.0[e]	42.0[e]	41.6
NPE10	1%	99.0[e]	26.6[c]	26.5
NPE18	1%	53.5[d]	14.2[b]	7.6
NPE30	1%	30.6[b]	12.4[a]	3.8

[a–e] Figures denoted by the same footnote letters do not differ significantly ($p = 0.05$; Duncan multiple range test)

Table 2. Relationship between penetration (diameter of inhibition zone of *Aspergillus niger* on agar under a treated epidermis disc, *Allium porrum*) and activity (against *Erisyphe graminis* on summer barley) of carbendazim and benomyl (mildew on summer barley; from Müller and Burth 1982)

Treatment	Penetration (mm)	Activity (%)
Carbendazim	8	2
Carbendazim + surfactant	51	96
Benomyl	24	21
Benomyl + surfactant	54	96

more pronounced than that on penetration. The cumulative effect, however, estimated by multiplying both factors, resulted in an important improvement of performance when the NPE10 concentration was higher than 0.1%. In addition, higher ethoxylated compounds (NPE18 and NPE30) were less efficient. The combined effects obtained by this theoretical approach are not always fully reflected by experimental results because of phytotoxic side effects which disturb the expected increase of uptake. This influence is very fungicide specific, as shown by Holloway et al. (1992), who obtained similar results for diclobutrazol but different uptake patterns for ethirimol. As observed by Müller and Burth (1992; Table 2), even closely related compounds, such as the benzimidazoles, carbendazim, and benomyl, penetrate in totally different ways both in the presence and absence of surfactants.

6.4 Influence of Phytotoxic Effects of Surfactants on Fungicide Performance

Phytotoxic effects attributable to the combined treatment with fungicides and surfactants can often result in decreased uptake due to the disturbance of essential processes in the plant, which in turn can restrict translocation of xylem-mobile compounds.

This has been demonstrated by Knoche et al. (1992), and some data from Vandenborre (1993) are presented in Table 3. The measurement of phytotoxicity is subject to much discussion (Coupland et al. 1989) because the parameters used often lead to conflicting conclusions and interpretations. Some observed effects do not actually reflect the vital influence of phytotoxic action but represent only nonsignificant side effects. The chlorophyll content of wheat leaves (Table 3) was lowered by fungal infection, but any difference between healthy and diseased plants disappeared after treatment with surfactant and/or fungicide. Although a decrease in chlorophyll content attributable to infection is

Table 3. Chlorophyll content (mg/g dried leaf) of wheat leaves and influence of *Oidium monilioides* infection (from Vandenborre 1993)

Treatment	Healthy plants	Diseased plants
Control	21	14
Surfactant (Atplus 469)	9	9
Propiconazole	8	6
Propiconazole + surfactant	13	10

Table 4. Phytotoxicity ratings according to different experimental parameters after treatment of *Cucumis sativus* in hydroculture (from Steurbaut et al. 1989)

Treatment	Phytotoxicity rating type				
	I	II	III	IV	V
None	A	A	A	A	A
Flutriafol	D	D	B	A	A
Flutriafol + Atplus 201	D	D	A	A	C
Propamocarb	C	B	A	B	B
Propamocarb + Atplus 201	C	D	A	B	C

I, Weight increase of plant; II, weight of root system III, uptake of nutrient solution; IV, nitrate uptake; V, ammonium uptake; A, 100%–75% of control; B, 75%–50%; C, 50%–25%; D, 25%.

expected (Agrios 1988; Scott and Smillie 1966), it was demonstrated by Daly (1976) that, especially during the initial phases of the infection, rate changes of photosynthesis can be regulated metabolically. In this way photosynthesis during the pathogenesis is not always related to the degree of infection and this is also a possible explanation for the different behavior of the tested fungicide (fenpropimorph, tridemorph, propiconazole, chlorothalonil) -surfactant combinations. Steurbaut et al. (1989) also found considerably diversity of phytotoxicity ratings according to different experimental parameters (Table 4).

6.5 Influence of Infection on Adjuvant Activity

That a fungal infection can play a role in the transport of a fungicide, and that surfactants can have an important role, can be deduced from the results presented in Table 5. Diseased plants may possibly allow greater penetration of the fungicide through the impaired leaf tissue. Although systemic fungicides can have a certain limited specificity, it is possible that a foliar fungicide-surfactant application to control a certain disease may result in increased susceptibility to other pathogens due to the disruption of the barrier function of the cuticle. For example, Marois et

Table 5. Influence in fungal attack on the translocation of three fungicides (from Humphreys and Ayres 1992)

Fungicide	Increase in translocation compared with healthy plants		
	Septoria nodoram	Erisyphe graminis	*Puccinia recondita*
Hexaconazole	× 1.2	× 0.7	× 1.1
Flutriafol	× 17.5	× 6.3	× 1.5
Dichlobutrazol	× 1.4	× 1.2	× 1.3

al. (1987) found substantial Botrytis infection on grapes after a treatment with surfactant in the absence of fungicide.

Although the translocation behavior of systemic fungicides is well documented, little attention has been directed to the influence of a local fungal infection on the movement and accumulation of these compounds. Humphreys and Ayres (1992) studied the effect of different fungi on the local accumulation of fungicide residues at the infection sites. Only flutriafol showed enhanced accumulation in the tissue infected by Septoria and mildew when compared with the corresponding tissue of healthy plants (Table 5). They also found very important differences between the fungi tested. The lack of accumulation of most compounds around rust lesions was unexpected since the infection would increase transpiration, potentially resulting in accumulation. Fungicide concentrations were higher in the plant tissue itself than in the fungal hyphae. The presence of surfactant probably enhanced the uptake of the AI through the weakened lesions of the infected tissue in the same way that it can affect the protective functions of the cuticular membranes of healthy plants. The only difference is the weakened defense mechanisms caused by the infection.

6.6 Influence of Surfactants on Resistance Development

The problem of fungicide resistance is complex, but one of the factors contributing to its promotion is prolonged exposure of the pathogen to (sub)lethal concentrations of the fungicide. In relation to this, the uptake and accumulation pattern of the AI in the leaf is very important. Several studies (Steurbaut et al. 1989, 1991; Holloway et al. 1992) have demonstrated, experimentally and by simulation, that the uptake of an AI in combination with surfactants is generally higher, faster, and more prolonged than in the absence of surfactant. This could result in conditions favorable for resistance development, but until now no indications of increased resistance have been reported in the literature. It is, however, less disputable that enhanced uptake of fungicides gives rise to higher concentrations, which have an improved lethal effect on fungi.

A remarkable observation by Farestra et al. (1992) was the cross-resistance of benzimidazole fungicides with a fosethyl formulation (Aliette). Because this resistance appeared only with the formulation and not with the AI, it was concluded that this activity is due to the surfactant(s) present in the formulation (200 g/kg). It was even concluded that investigations on the nature of these active adjuvants might result in the discovery of new compounds effective against gray mold.

6.7 Influence of Fungal Cell Wall Constitution and Location of the Site of Action

A fungitoxic compound must penetrate through the fungal cell in order to reach its site of action. The penetration processes are governed by the same mechanisms as those involved with plant cuticles, although some particular aspects related with the fungal structure are involved. First of all, there is a great diversity among fungi. The form and properties of the mycelium of different fungi is at least as diverse as the differences between plants, where it has been shown that a whole range of physicochemical (e.g., wax constitution) and morphological (e.g., hairiness, roughness, and venation) properties can influence contact phenomena and the penetration potential of the AI. Surfactants can interfere with fungal penetration processes in the same way as they do on and in leaf surfaces. Also the sensitivity and the vital state of some fungal forms during growth and development may be totally different (Spott and Cervantes 1987; Stanghellini and Tomlinson 1987).

Extrapolation from and comparisons with different experimental systems is sometimes very debatable. Aerial treatments of mycelia can be very problematic due to difficulties in wetting as compared with in vitro testing on contaminated nutrient agar plates, where uptake by the sensitive hyphal organs is much easier (Amer et al. 1992). Favorable disease control from some fungicide treatments can be seen in the perspective of a temporary disorder of the leaf surface structure, whereby the surfactant action can be expected to be more effective when progress of the disease by vigorous germ tube development is taking place.

Surfactants interfere not only with the defense membrane function of the leaf surface but also with the infection stages of fungi. The observations of Crookes and Laing (1988), studying the influence of surfactants on the rust infection of wheat, are in this respect very interesting. They found that germ tube extension on control plants is at right angles to the leaf venation with only limited branching. In addition, strong adhesion to the leaf and formation of typical infection structures at stomata were observed. Surfactant-treated plants gave rise to prolific and nondirectional germ tube growth with profuse branching and a random production of ineffective infection structures. It therefore seems that surfactants can adversely affect the preinfection stimulation responses, resulting in less successful infection. The mechanisms involved are unknown, but it is possible that the surfactant deposits on the leaf surface can interfere with adhesion of the germ tubes. Although the comparison is not direct, there is for this phenomenon some kind of resemblance with a phytoalexin-type reaction and an elicitor effect. More importantly, however, these observations demonstrate that even sublethal doses of surfactants can interfere with the fungal infection processes.

6.8 Influence of Treatment Type

The importance of curative versus protective treatments in relation to fungicide efficiency is connected with the potential for penetration of the AI into the tissue which may be modified by surfactant treatment. Surfactants can enhance fungicide penetration but may also have a secondary effect due to disturbance of the wax and cuticular layers. However, it is not always obvious that this disorganization gives rise to an improvement because a change from a rather rough and difficult-to-wet cuticular surface to a smooth surface does not always improve wetting, spreading, and/or penetration (Watanabe and Yamaguchi 1992). Also the timing of applications can be very critical because the initial disruptive effect can be repaired quickly by plant recuperative processes, especially in young and actively growing plant tissues. Confusing results may thus be obtained due to the unpredictable disease progress as a function of time. For example, Engelke and Ansermino (1991) found excellent control of cabbage downy mildew by a fungicide-surfactant combination in the early stages of infection, but after 2 weeks this controlling effect became greatly reduced. The disease eventually became more severe than the corresponding fungicide application without surfactant.

6.9 Conclusions

Applied and fundamental research on the influence of surfactants at the first steps of their potential activity are well-studied, for example, improved droplet formation and deposition, increased target wetting, and coverage. Also, some progress has been made in understanding the mechanisms of enhanced cuticular and translaminar transport and translocation. On the other hand, the mode of action of some fungicidal active materials, such as the inhibition of ergosterol biosynthesis, is well understood. Nevertheless, there are still some missing links on both sides in relation to the complex transfer of the AI from the plant to the fungicidal site of action. This involves delicate equilibria between fungicide–surfactant combinations and plant-fungus systems. Because of the high specificity of such relationships it would appear that there are no general rules for fungicides, unlike those for many herbicides. Only specific, tailor-made formulations for each disease-control situation are appropriate. It is therefore the task of the fungicide formulator to optimize a particular application for a well-defined disease on a certain crop and to consider appropriate surfactants as formulation additives for enhancing activity.

References

Agrios GN (1988) Plant pathology. Academic, San Diego

Amer MA, Poppe J, Hoorne H (1992) Adjuvants for enhanced fungicide efficacy against Botrytis cinerea and Septoria apiicola. Medelingen van de Faculteit Landbouwwetenschappen Rijksuniversiteit Gent 57:127–138

Baker EA, Hayes AL, Butler RC (1992) Physicochemical properties of agrochemicals: their effects on foliar penetration. Pestic Sci 34:167–182

Bauer H, Schönherr J (1992) Determination of mobilities of organic components in plant cuticles and correlation with molar volumes. Pestic Sci 35:1–11

Chaumat E, Chamel A, Taillandier G, Tissut M (1992) Quantitative relationships between structure and penetration of phenylurea herbicides through isolated plant cuticles. Chemosphere 24:189–200

Chamel A, Gambonnet B, Coret J (1992) Effect of two ethoxylated nonylphenols on sorption and penetration of isoproturon through isolated plant cuticles. Plant Physiol Biochem 30:713–721

Coupland D, Zabkiewiez JA, Ede FJ (1989) Evaluation of the techniques used to determine surfactant toxicity. Ann Appl Biol 115:147–156

Crookes CA, Laing MD (1988) Pre-infection stages of wheat leaf rust. Proc Electron Microsc Soc South Afr 18:103–104

Daly JM (1970) The carbon balance of diseased plants: changes in respiration, photosynthesis and translocation. In: Heitefuss R, Williams PH (eds) Physiological plant pathology. Springer, Berlin Heidelberg New York, pp 450-475

Engelke RD, Ansermino SD (1991) The effectiveness of the experimental product S151 relative to other stabilised products, as well the effect of the experimental wetter, EN600 together with S151, in controlling downy mildew (Peronospora parasitica) on cabbage (Brassica oleraceae var. capitata cv 'Greenstone'). University of Natal, Pietermaritzburg, p 11

Farestra F, Pollastro S, Sansiviero F (1992) Activity of adjuvants in Aliette against Botryotinia fuckeliana (Botrytis cinerea) and their cross-resistance with benzimidazole fungicides. Pestic Sci 35:171–174

Foy CL, Chow PNP, Grant A (1992) Appendix. In: Foy CL (ed) Adjuvants for agrochemicals. CRC Press, Boca Raton, pp 691–715

Holloway PJ, Wong WW, Partridge HJ, Seaman D, Perry RB (1992) Effects of some nonionic polyoxyethylene surfactants on uptake of ethirimol and diclobutrazol from suspension formulation applied to wheat leaves. Pestic Sci 34:109–118

Humphreys SP, Ayres PG (1992) Effects of fungal infection sites on the distribution of fungicides. Pestic Sci 34:185–186

Knoche M, Noga G, Lenz F (1992) Surfactant – induced phytotoxicity: evidence for interaction with epicuticular wax fine structure. Crop Protect 11:51–56

Marois JJ, Bledsoe AM, Bostock RM, Gubler WD (1987) Effects of spray adjuvants on development of Botrytis cinerea on Vitis vinifera berries. Phytopathology 77:1148–1152

Müller R, Burth U (1982) Benzimidazol-Tensid – Formulierung mit Wirksamkeit bei Benomylresistenz. Abhandl Akad Wissensch DDR 1:315–319

Scott KJ, Smillie RM (1966) Metabolic regulation in diseased leaves. I. The respiratory rise in barley leaves infected with powdery mildew. Plant Physiol 41:289–297

Spott RA, Cervantes IA (1987) Effects of the nonionic surfactant Ag-98 on three decay fungi of Anjou pear. Plant Dis 77:112–114

Stanghellini ME, Tomlinson JA (1987) Inhibitory and lytic effects on a nonionic surfactant on various asexual stages in the life cycle of pythium and phytophthora species. Phytopathology 77:112–114

Steurbaut W (1993) Adjuvants for use with foliar fungicides. Pestic Sci 38:85–91

Steurbaut W, Melkebeke G, Dejonckheere W (1989) The influence of nonionic surfactants on the penetration and transport of systemic fungicides in plants. In: Chow PNP, Grant CA, Hinshalwood AM, Simundsson E (eds) Adjuvants for agrochemicals. CRC Press, Boca Raton, pp 94–103

Steurbaut W, Megahed HS, Van Roey G, Melkebeke T, Dejonckheere W (1991) Influence of surfactant-oil combinations on the activity of foliar-applied fungicides. In: Foy CL (ed) Adjuvants for agrochemicals. CRC Press, Boca Raton, pp 623–635

Stock D, Edgerton BM, Gaskin RE, Holloway PJ (1992) Surfactant-enhanced foliar uptake of some organic compounds: interaction with two model polyoxyethylene aliphatic alcohols. Pestic Sci 34:233–244

Vandenborre R (1993) Evaluation of adjuvants for the chemical control of oidium monilioides nees on wheat. PhD thesis, University of Gent, Belgium, p 70

Watanabe T, Yamaguchi I (1992) Studies on wetting phenomena on plant leaf surfaces. 3: retention model for droplets on solid surfaces. Pestic Sci 34:273–279

7 Evaluation of Adjuvant Modes of Action: A Theoretical Approach

P. J. Holloway

7.1 Introduction

Many products are available commercially as tank-mix or built-in adjuvants for improving the efficiency of various stages in the application of pesticides to foliage (recent reviews by Chapman and Mason 1993; Foy 1993; Holloway 1993). Because most adjuvants are usually sold for a specific purpose it is often taken for granted that their modes of action are well understood. However, this is the exception rather than the rule, since in many cases successful products have evolved mainly from extensive but empirical screening, especially for beneficial effects on the biological activity of pesticides. From the point of view of marketing and labelling of adjuvants, it is important for the end-user to have basic information on how a particular adjuvant works so that he can assess its relative merits and usefulness for a crop protection programme. For this to be achieved reliable quantitative methods for evaluating performance and activity are required both in the laboratory and in the field. In the past such an approach has not been given sufficient priority by the industry, which as a result has been tarnished by a 'snake oil' and 'spray-and-pray' image.

The main objective of this short contribution is therefore to suggest some methodology that is appropriate for monitoring specific adjuvant effects and, in turn, for identifying possible modes of action, as opposed to precise mechanisms. This should always be considered in relation to the site of action required which may be (a) in the spray tank only, (b) in the spray droplets, (c) in the spray deposits or (d) inside the target after penetration.

7.2 Adjuvant Types

Spray additives can be classified either according to their chemical composition or to their usage (Holloway 1993). The main chemical classes, in order of commercial importance, are: surfactants, oils (mineral or vegetable), polymers, polymer-forming compounds and inorganic salts. For the purposes of evaluation, however, it is more appropriate to consider usage categories because these usually refer to specific modes of action. Thus, criteria for assessing the following typical label designations need to be considered:

- Activator/penetrator
- Antidrift agent
- Antievaporant
- Antifoam agent
- Buffering agent
- Compatibility agent
- Deposition agent
- Safener
- Sequestering agent
- Spreader
- Sticker
- Synergist
- UV-screening agent
- Wetter

Although these categories are intended to describe the primary mode of action, it should be noted that secondary effects can also be produced, which may or may not be beneficial to the overall pesticide response.

7.3 State-of-Art Methodology

At the present time there are no official standardisation procedures specifically for agrochemical adjuvants (Roberts 1992). Consequently, it is left to the vendor to provide scientific evidence in support of the label description; in practice this varies considerably in the amount of detail provided. Bioassay (see Kudske and Mathiassen, this volume) in the laboratory and field is the mainstay of adjuvant testing.

Other methodology, however, is also available for measuring the salient properties of adjuvants either in bulk solution, from small droplets delivered from either a microsyringe or droplet generator, or using droplets produced by a sprayer. Adjuvants can be tested in the presence or absence of the pesticidal active ingredient and its coformulants. Understandably, most emphasis has been given to surfactants, for which some methodology already exists from other applications, such as textiles and detergents (Cox 1989; Roberts 1992). For example, equilibrium and dynamic surface tension measurements are relevant to the performance of antifoam agents, some deposition agents, spreaders and wetters. Contact angles and deposit areas formed on artificial or leaf surfaces provide additional documentation of the relative efficacy of spreaders.

The efficiency of wetters and deposition agents can be quantified from track sprayer experiments using solutions of dyes or the pesticide itself, provided that a sensitive analytical method is available; such information rarely appears in the technical literature from companies about such adjuvants. More sophisticated equipment for in-flight droplet sizing is needed to assess the potential of antidrift agents. Rainfastness of the deposit is the criterion for stickers. To demonstrate the mode of action of an adjuvant as an activator/penetrator it is essential to employ radiochemical methods in conjunction with labelled pesticides (see Gauvrit, Stock and Davies, this volume); such techniques are also useful for identifying safener and synergist activity in vivo and in vitro. If radiolabelled adjuvant is available, even more precise information can be obtained about its site and mechanism of action (see Gauvrit, this volume). Methods are also available to study adjuvant action at the cellular level (see, for example, Røyneberg et al. 1992).

If specific adjuvant properties can be related to bioassay data, this should reduce the need for expensive and time-consuming empirical testing of adjuvant-pesticide combinations. In this context it should be noted that residual, contact and systemic pesticides will almost certainly have different adjuvant requirements in order to improve their biological performance.

7.4 Future Prospects

It is forecast that adjuvant usage will provide an important growth market, especially with the increasing emphasis on the need to reduce pesticide application rates in agricultural practice. In the United States, for example, the projected value of adjuvant and coformulant sales for 1997 is $365 million, an increase of 24% from 1992 (Anonymous 1993). However, it is felt by many that development of the market is hampered by misconceptions, incomplete scientific knowledge and extravagant claims for efficacy. The situation is compounded by the lack of regulatory harmonisation, especially in Europe, where Denmark has no approval scheme, the United Kingdom and Ireland require only safety data for registration, whilst France and Greece require both safety and efficacy information.

It is generally agreed that there will be new legislation and more stringent registration requirements for adjuvants in the future. In anticipation of these, the following are suggested as possible criteria for evaluation and acceptability purposes; they could be useful for compiling a broad database for existing and new products. Thus, does the adjuvant affect:

- Spray atomisation?
- Droplet evaporation during flight?
- Droplet drift?
- Droplet deposition and retention?
- Droplet spreading?
- Overall target coverage?
- Rate of droplet evaporation?
- Physical form of spray deposit?
- Moisture content of spray deposit?
- Rate and amount of uptake of pesticides and which ones in particular?
- Rate and amount of translocation of systemic pesticides and which ones in particular?
- Rate of metabolism of pesticides and which ones in particular?

Other pertinent questions which should be asked are:

- On what plant species is the adjuvant most beneficial?
- At what concentration is the adjuvant most effective?
- Are there any problems with intrinsic phytotoxicity?
- Is the product compatible with other formulants and adjuvants?
- How safe is the adjuvant to use?
- What is the toxicity to fish, mammals and beneficial invertebrates?
- How rapidly is the product biodegraded in soil and plants?
- What are the potential cost benefits?

Standards for adjuvant efficacy and mode of action should not only provide a more rational and scientific footing for the industry but also ensure that the customer is getting what he has paid for. Surely, it is not unreasonable to expect, for example, that an adjuvant sold as a wetter actually improves spray retention. Apart from efficacy, toxicological, environmental and cost considerations are likely to be major factors governing adjuvant acceptance in the future. Agrochemical companies prefer to sell all-in-one products because the pesticide, coformulants and adjuvants can be submitted for registration purposes at the same time.

Acknowledgements. The research on adjuvants at Long Ashton Research Station is funded by a commission from the Ministry of Agriculture, Fisheries and Food. The author thanks Mrs J. Hynam for production of the manuscript.

References

Anonymous (1993) Adjuvants, a US growth market. AGROW 181:11

Chapman PJ, Mason RD (1993) British and European Community regulations and registration requirements for non-pesticidal co-formulants in pesticides and for adjuvants. Pestic Sci 37:167–171

Cox MF (1989) Effect of alkyl carbon chain length and ethylene oxide content on the performance of linear alcohol ethoxylates. J Am Oil Chem Soc 66:367–374

Foy CL (1993) Progress and developments in adjuvant use since 1989 in the USA. Pestic Sci 38:65–76

Holloway PJ (1993) Adjuvants for agrochemicals: why do we need them? Medelingen van de Faculteit Landbouwwetenschappen Rijksuniversiteit Gent 58/2a:125–140

Roberts JR (1992) A review of the methodology employed in the laboratory evaluation of spray adjuvants. In: Foy CL (ed) Adjuvants for agrichemicals. CRC, Boca Raton, pp 503–512

Røyneberg T, Balke NE, Lund-Høie K (1992) Effects of adjuvants and temperature on glyphosate absorption by cultured cells of velvetleaf (Abutilon treophrasti Medic). Weed Res 32:419–428

8 Methodology for the Study of Spray Application and Biological Efficacy of Herbicides on Pot-Grown Plants

P. Kudsk and S. K. Mathiassen

8.1 Introduction

Numerous aspects are taken into consideration when formulating herbicides. Although problems such as storage stability, ease of mixing with water, and operator exposure when handling the product must be solved, the most important single aspect is to optimize biological efficacy.

The selection of formulations to optimize activity and selectivity normally takes place within chemical companies, and their exact composition is proprietary information. Off-patent herbicides in particular may be available as various formulations, and it is therefore also of interest outside the chemical companies to assess activity of formulations.

With regard to biological efficacy, adjuvants are the most important formulation constituents. Besides adjuvants present in the formulations, adjuvants can also be added to the spray solution by the end user along with the herbicide. The effect of adjuvants on herbicide efficacy has been known since the introduction of the first organic herbicides more than 60 years ago (Foy 1989). Bioassays using plants grown in pots in glasshouses have been a commonly used technique to assess formulations and adjuvants over this time. Such whole-plant bioasssays represent what have been termed a supercomplex system (Fedtke 1982), and as such they often cannot provide any explanation of observed differences in biological activity. However, they do come close to the field situation, without the costs of field experiments and whole plant bioassays, and therefore continue to be a widely used method in formulation and adjuvant research and should consequently be developed further.

Following a short state-of-the-art review, we present results from our research on the assessment of formulations and adjuvants, influence of adjuvant dose, and adjuvants as modifiers under adverse application conditions. Some of the results presented in this paper have been published previously (Kudsk and Streibig 1993). As appropiate statistical analyses are crucial for interpreting results from whole-plant bioassays, a section is devoted to describing the nonlinear regression techniques used in our research work.

8.2 State-of-the-Art

The number of publications on adjuvant studies using whole-plant bioassays is very large. Early research focused on herbicides such as 2,4-D (e.g., Marth et al. 1945; Mitchell and Hamner 1944) and later on diuron and atrazine (e.g., Bayer and Drever 1965; Dexter et al. 1966; McWorter 1963a), two herbicides otherwise possessing little or no foliar activity. Later, numerous papers were published on adjuvants for glyphosate (e.g., Turner and Loader 1980; Wyrill and Burnside 1977), and recently much work has been directed at identifying effective adjuvants for selective grass weed herbicides of the aryloxyphenoxypropionate and cyclohexadione classes (e.g., Keeney et al. 1988; Manthey et al. 1989).

When reviewing the literature, it is remarkable that many of the studies have essentially tested more or less randomly selected adjuvants at a single dose rate. Only very few studies have been undertaken to study the relationship between the efficacy of herbicide-adjuvant mixtures and the chemical structure and/or physicochemical properties of the adjuvants and the influence of adjuvant dose. In this context the work of Jansen and coworkers (Jansen 1964, 1965, 1973; Jansen et al. 1961) and Smith et al. (1966) with a series of structurally related surfactants is still exceptional.

In most studies efficacy has been assessed either by determining fresh or dry weights or by visual assessment of injury. In a few, such as the work by Jansen (Jansen 1964, 1965; Jansen et al. 1961) visual assessments and height and weight measurements were combined in a rather obscure activity index. In general, determination of biomass in terms of fresh or dry weights should be used as this is an objective measurement and is probably most relevant to the crop-weed competiton situation in the field. Visual assessments, however, can overcome problems with a heterogenous plant material which could otherwise make it difficult to draw unequivocal conclusions from fresh and dry weights.

The majority of experiments have been conducted under glasshouse conditions. In our work we have observed that responses to an adjuvant may differ significantly between glasshouse and outdoor-grown plants. Whereas the leaf surface of outdoor plants is damaged, for example, by wind abrasion and natural rainfall, the leaf surface of glasshouse-grown plants is normally intact. Furthermore, the growth habit of glasshouse-grown plants often differs from that of outdoor grown-plants which, on the other hand, are very similar to plants in the field. For example, glasshouse-grown grass species are often more erect than those grown outdoors. In particular, with spray solutions with a high surface tension applied to difficult-to-wet species these differences often result in much more pronounced responses to adjuvants on glasshouse than on outdoor-grown plants because spray retention becomes very critical. This emphasizes that one should be careful when interpreting results from glasshouse experiments, and we therefore conduct the majority of our adjuvant studies on outdoor-grown plants.

Another striking feature is that many of the published works have examined only a few herbicide dose rates, very often only one. The selection of dose rates becomes extremely important when only one or two doses are included because one cannot expect to see significant effects of adjuvants at high dose rates as plants are killed whether adjuvants are included or not. Even if a dose rate is carefully selected to yield a 50% growth reduction for the herbicide alone, addition of adjuvants often results in complete kill, rendering it impossible to distinguish between different adjuvants. Nevertheless, even if one succeeds in selecting a proper herbicide dose rate, conclusions drawn from such a study are specific for this dose rate and are not general. Consequently, several dose rates must be included in an experiment to separate adjuvant effects from dose-response effects. This calls for the use of nonlinear regression techniques in adjuvant studies instead of the traditionally applied analyses of variance.

8.3 Concept of Parallel Dose-Response Curves

While the effect of a herbicide in most cases can be described by a logistic dose-response curve, an adjuvant or a formulation blank is virtually without any effect on plant growth at normal dose rates. Figure 1A shows typical dose-response curves for a herbicide and an adjuvant or formulation blank. Consequently, we can consider all constituents other than the herbicide molecule biologically inactive when applying formulated herbicides or mixtures of herbicides and adjuvants. Although biologically inactive when applied alone, adjuvants and other formulation constituents can influence herbicide efficacy by affecting spray retention both quantitatively and qualitatively, as well as herbicide uptake and translocation, and thus the amount of herbicide reaching its site(s) of action. On the other hand, formulations and adjuvants are assumed not to influence the mode of action of the herbicide, and consequently it is reasonable to assume that the dose-response curves for different formulations of the same herbicide or a herbicide applied alone and in admixture with an adjuvant must be parallel (Fig. 1B). In other words, the applied dose, or what could be called the apparent dose, is the same, but the effective dose, i.e., the amount reaching the site of action depends on formulation or adjuvant. The effect of formulations or adjuvants is reflected in the magnitude of the horizontal displacement of the dose-response curve, i.e., by a horizontal assessment instead of a

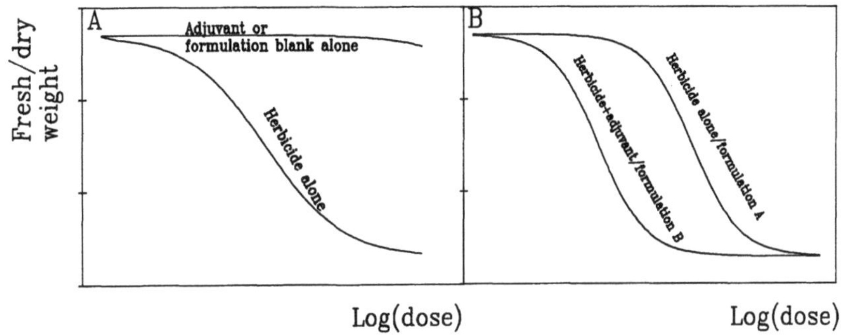

Fig. 1. Schematic dose-response curve of a herbicide and an adjuvant/formulation blank (**A**) and a herbicide applied alone and in combination with an adjuvant or two formulations of the same herbicide (**B**). (Reprinted with permission from Kudsk and Streibig 1993; copyright CRC Press, Boca Raton)

vertical assessment as in an analysis of variance. While the concept of parallel dose-response curves has been widely used in pharmaceutical research when comparing different preparations of drugs (Finney 1978), it has rarely been used in herbicide research, although the principles of testing drug preparations and herbicide formulations are similar.

The dose-response curve for a standard preparation, for example, a commercial formulation, can be described by the equation of Finney (1979):

$$U = \frac{D - C}{1 + \exp\{-2\,[(a + b)]\,\log\,(z)]\}} + C,\ b < 0 \qquad (1)$$

where U is the plant response (fresh or dry weight), z is the herbicide dose, D and C denote the upper and lower limits at zero and high doses, and a and b are parameters describing the horizontal location and the slope around the point of inflection (ED_{50}). A very important property of Eq. 1 is that the four parameters can be interpreted biologically. Assuming parallel dose-response curves, the dose-response curve for a test preparation, for example, a test formulation or the commercial formulation with an additional adjuvant, can then be described by:

$$U = \frac{D - C}{1 + \exp\{-2\,[a + b \times \log\,(Rz)]\}} + C,\ b < 0 \qquad (2)$$

The only difference between Eqs.1 and 2 is the parameter R; this refers to the relative potency and expresses the ratio of the dose rates of the standard and test preparations resulting in the same effect. As the dose-response curves are parallel, R is constant and independent of dose rates and response level. By applying the concept of parallel dose-response curves it is possible to overcome the problem mentioned in Sect. 8.2 of separating adjuvant and dose-response effects. Therefore, conclusions about the intrinsic effect of formulations and adjuvant from studies applying parallel dose-response curves are general and not specific. Furthermore, the effect of formulations and adjuvants are expressed quantitatively and not only qualitatively, as with an ordinary analysis of variance. If the costs of active ingredient and formulations or herbicide and adjuvant are known, it is possible to determine the most cost-effective formulation or herbicide-adjuvant combinations. An example is given in Sect. 8.4.2.

ED_{50} values can be calculated from antilog ($-a/b$). Hence, Eqs. 1 and 2 can be reparameterized to comprise ED_{50} as a parameter resulting in Eqs. 3 and 4:

$$U = \frac{D - C}{1 + \exp\,[2\,b\,(\log\,(ED_{50}) - \log\,(z)]} + C,\ b < 0 \qquad (3)$$

$$U = \frac{D - C}{1 + \exp\,\{2\,b\,[\log\,(ED_{50}) - \log\,(Rz)]\}} + C,\ b < 0 \qquad (4)$$

Any response level can be used as a parameter in Eqs. 3 and 4, i.e., ED_{50} can be replaced, for example, by ED_{90} which from a practical point of view is more relevant than ED_{50}. A more detailed description on dose-response curves and statistical methods can be found elsewhere (Streibig et al. 1993).

The following sections present a number of examples to illustrate how nonlinear regressions and, in particular, the concept of parallel dose-response curves can be applied in herbicide bioasssays with whole plants. In these experiments responses were transformed by using a transform-both-side technique to stabilize the variance (Rudemo et al. 1989). A power transformation of either 0 (logarithmic transformation) or 0.2 was found to be appropiate. Results were analyzed by an analysis of variance and by nonlinear regressions assuming parallel and nonparallel dose-response curves, respectively. A test for lack of fit comparing nonparallel regression and analysis of variance revealed whether the logistic model could be accepted. If not rejected, a further test for lack of fit comparing parallel and nonparallel regressions showed whether the hypothesis of parallel dose-response curves was appropiate. The latter test for lack of fit is given in the tables; for the sake of clarity, only the estimated relative potencies and not the other parameter estimates are shown.

8.4 Assessment of Efficacy of Formulations and Adjuvants

8.4.1 Formulations and Adjuvants

Relative to the costs of developing a herbicide, the costs of formulating a herbicide are usually insignificant. The number of papers showing the benefits of adding adjuvants to commercial herbicides may therefore seem surprising. Weed flora, climatic conditions, and application techniques vary considerably around the world, and it could therefore be anticipated that the relatively few formulations available for each herbicide would not be optimal for every purpose.

In addition to improving activity, changes in formulation of commercial herbicides are often required because of the increased public concern about, for example, organic solvents. Fluazifop-*p*-butyl is a classical example of a herbicide which has been subjected to several changes in formulations. Ten years ago it was marketed as an emulsifier concentrate (EC) formulation in Denmark, but some years ago an oil-in-water (EW) formulation was introduced to reduce the amount of organic solvents in the formulation. Recently, a wettable powder (WG) formulation, a type of formulation which has become popular due to the ease of handling and reduced risk of operator exposure, was introduced. The activity of the EW and WG formulations has been compared on barley (*Hordeum vulgare*) and Italian rye-grass (*Lolium multiflorum*). The formulations were applied in admixture with the recommended adjuvant for the EW formulation, a C_9/C_{11} primary alcohol polyethoxylate surfactant with an average of 6 ethylene oxide (E) units (AE6), and TF8035, an experimental mineral oil adjuvant of unknown composition. The efficacy of the two formulations was similar irrespective of adju-

Table 1. Estimated relative potencies of an EW and WG formulation of fluazifop-*p*-butyl in admixture with two adjuvants on barley and Italian rye-grass

	Formulation	Adjuvant	Relative potency	95% CI[x]
Barley[a]				
	EW	0.1% AE6	1.0	
	EW	0.5% TF8035	1.13	1.03–1.22
	WG	0.1% AE6	0.95	0.87–1.03
	WG	0.5% TF8035	1.08	0.99–1.16
Italian rye-grass[b]				
	EW	0.1% AE6	1.0	
	EW	0.5% TF8035	1.54	1.32–1.77
	WG	0.1% AE6	1.01	0.86–1.15
	WG	0.5% TF8035	1.58	1.36–1.81

[x]CI, confidence interval
[a]Test for lack of fit: $F_{3,51} = 1.47$
[b]Test for lack of fit: $F_{3,65} = 0.92$

vant (Table 1). Only minor, but significant, differences were observed between the two adjuvants on barley, whereas on Italian rye-grass both formulations performed markedly better in combination with TF8035 than in combination with the surfactant. A relative potency of 1.54, as found on Italian rye-grass with the EW formulation in admixture with TF8035, indicates that by shifting from the nonionic surfactant to the mineral oil adjuvant the dose of fluazifop-*p*-butyl can be reduced to 65% (1/1.54) of the dose required in admixture with the surfactant without loss of efficacy. This example clearly illustrates how the concept of parallel dose-response curves can be applied to assess the influence of both formulation and adjuvants.

A research project that we began this year is studying the influence of a series of nonionic surfactants at various concentrations on herbicide efficacy to determine the physicochemical factors that influence the surfactant-induced enhancement of biological activity. Several experiments with sulfonylurea herbicides have been conducted, and Fig. 2 presents some of the results obtained with tribenuron-methyl. The herbicide was applied at five dose rates alone and in admixture with four

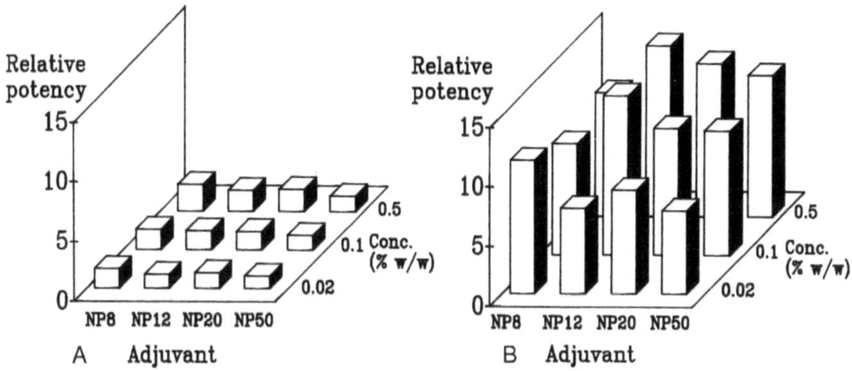

Fig. 2. Estimated relative potencies of tribenuron methyl in admixture with NPE8, NPE12, NPE20, and NPE50 at 0.02, 0.1 and 0.5% (w/w) on *Lamium hybridum* (**A**) and *Chenopodium album* (**B**). No surfactant, 1.0

nonylphenol polyethoxylates with an average of 8 (NPE8), 12 (NPE12), 20 (NPE20) and 50 (NPE50) E0 units at three concentrations, 0.02, 0.1, and 0.5% (w/w). Four plant species were included, and the results for *Lamium hybridum* and *Chenopodium album* are shown in Fig. 2. Within each plant species, 13 dose-response curves were fitted simultaneously to the fresh weights with the no surfactant treatment as standard (relative potency, 1). Only minor effects were found according to EO content. Tribenuron-methyl is a compound of intermediate polarity and the lack of effect of EO content on herbicide efficacy is in agreement with the predictive uptake model suggested by Stock et al. (1993). In contrast, with the exception of NPE8, herbicide activity increased with increasing surfactant concentration on both plant species. The most marked difference was between plant species. On *L. hybridum*, the relative potency of the herbicide-surfactant combinations varied between 1.05 and 2.22, whereas on *C. album* the relative potencies were 7.00–14.36. This difference can probably be attributed to a more pronounced enhancement of spray retention by the surfactants on *C. album*, a difficult-to-wet species, than on *L. hybridium*, an easy-to-wet species. Differences in the ability of the surfactants to promote uptake on the two plant species may also partly explain the observed differences. More detailed studies are in progress to explain the different response of the two plant species.

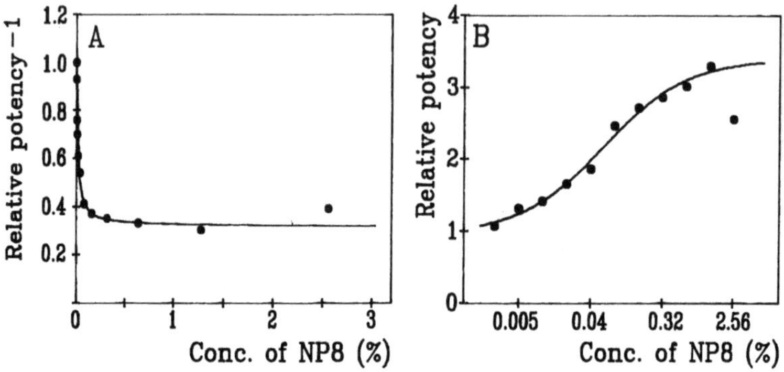

Fig. 3. Dose-response curves and observed mean fresh weights of *Poa annua* (**A**) and oilseed rape (**B**) after application of metribuzin alone and in admixture with 1% of vegetable oil adjuvant consisting of 90% rapeseed oil and 10% emulsifiers

Previous studies examining the influence of series of structurally related surfactants employed only one herbicide dose rate (Jansen 1964; Smith et al. 1966; Becher and Becher 1969). Although other dose rates would probably not have changed the ranking of the surfactants, it is likely that the magnitude of the differences would have been different, i.e., conclusions based on one dose rate are specific and not general. In contrast, the relative potencies shown in Fig. 2 are based on the horizontal displacement of the parallel dose-response curves, and consequently the intrinsic effects of the members of the NP series of surfactant are independent of effect level and herbicide dose rate.

From a theoretical point of view it could be argued that parallel dose-response curves are not an appropiate assumption in all situations. If, for example, a commercial herbicide is tested alone and in admixture with an adjuvant, it can be anticipated that the effect of the adjuvant on herbicide activity will be more pronounced at low herbicide doses due to the lower concentration of any built-in adjuvant. Our experience shows, however, that the hypothesis of parallel dose-response curves is seldom rejected in formulation and adjuvant studies, indicating that it is an appropiate assumption. Nevertheless, we found in one experiment a marked deviation from parallelism. In a series of experiments the influence of a rape seed oil adjuvant on the foliar activity of metribuzin was examined. On most plant species, e.g., *Galium aparine* and *Poa annua*, the vegetable oil adjuvant promoted metribuzin activity and enhancement could be described as a horizontal displacement of the dose-response curve for metribuzin alone (Fig. 3A). On oilseed rape (*Brassica napus*), however, dose-response curves were not parallel (Fig. 3B), and the hypothesis of parallel dose-response curves was of course rejected by the test for lack of fit. At low doses the adjuvant enhanced metribuzin activity while at higher doses it antagonized it. On oilseed rape, a difficult-to-wet species, the adjuvant may affect both retention and uptake. Improved retention is the most likely effect of the adjuvant, whereas the results at the high doses strongly indicate that the adjuvant restricted uptake. One explanation of the observed phenomenon on oilseed rape could be that the two oppositely directed effects make different contributions at low and high doses. Another possibility is a dose-dependent interaction between metribuzin and the adjuvant on uptake. Research is in progress to determine whether this effect is specific to this adjuvant or is general. If dose-response curves are not parallel, the relative potency depends on dose and response level. It is therefore necessary to calculate the relative potency at more response levels to assess the effect of an adjuvant on herbicide activity.

8.4.2 Adjuvant Dose and Herbicide Efficacy

Most studies employing more than one adjuvant dose have reported a progressive enhancement of herbicide activity with increasing adjuvant dose (e.g., Bayer and Drever 1965; Jansen 1965, 1973; McWorther 1963a,b; Smith et al. 1966). Jansen et al. (1961), examining the effect of 63 surfactants on the efficacy of four herbicides on two plant species, also found examples of progressive reduction of herbicide efficacy with increasing adjuvant doses. However, none of these studies attempted to describe the relationship between herbicide efficacy and adjuvant dose.

Several reference models are available for describing the joint action of herbicides (Green and Streibig 1993). Because adjuvants can be considered biologically inert, we have with herbicide-adjuvant mixtures a limiting case of a reference model for assessing the joint action of herbicides, i.e., they call for the same reference models as do herbicide mixtures. Hewlett (1979) has suggested that the effect of a synergist, as adjuvants are in most cases, can be described by a rectangular hyperbola. If z_{herb} is the dose of the herbicide and z_{adj} the dose of the adjuvant, the relationship (m_1, m_2, and m_3 are parameters with positive values) can be described by:

$$z_{herb} = \frac{m_1 + m_2}{m_3 + z_{adj}} \tag{5}$$

The relative potency expresses the ratio between herbicide doses giving the same effect and the ED_{50} or ED_{90} values of herbicide-adjuvant mixture are calculated by dividing the corresponding values of the herbicide without adjuvant by the relative potency. Consequently, instead of using ED_{50} or ED_{90} values, as Hewlett did, z_{herb} in Eq. 5 can be replaced by $1/R$.

Based on the estimated m parameters, the joint action ratio (JAR), expressing the relative potency of the herbicide at indefinitely large adjuvant doses, can be calculated as:

$$\mathrm{JAR} = 1 + \frac{m_2}{m_1\, m_2} \tag{6}$$

The asymptotes of the rectangular hyperbola are m_1 and $-m_3$, but otherwise the m parameters have no biological significance. We therefore conducted a number of experiments including up to 11 adjuvant doses to examine whether the logistic model can be used to describe the effect of increasing adjuvant doses. The relative potencies were estimated by fitting 11 parallel dose-response curves simultaneously to dry weights (no surfactant = 1) and then fitting the following logistic curve to the estimated relative potencies:

$$R = \frac{D - C}{1 + \exp\{-2\,[(a + b)\log(z_{adj})]\}}\, C, b > 0 \tag{7}$$

The parameters denote the same as in Eq. 1, the only difference being $b > 0$. The sigmoid curve implies, as Hewlett's rectangular hyperbola, a maximum potency at indefinitely high doses.

Figure 4 presents the estimated curves using Hewlett's rectangular hyperbola (A) and the logistic model in Eq. 7 (B) on the results of an experiment on barley with fluazifop-*p*-butyl and 11 doses of a nonylphenol-polyethoxylate surfactant with an average of 8 E units (NPE8).

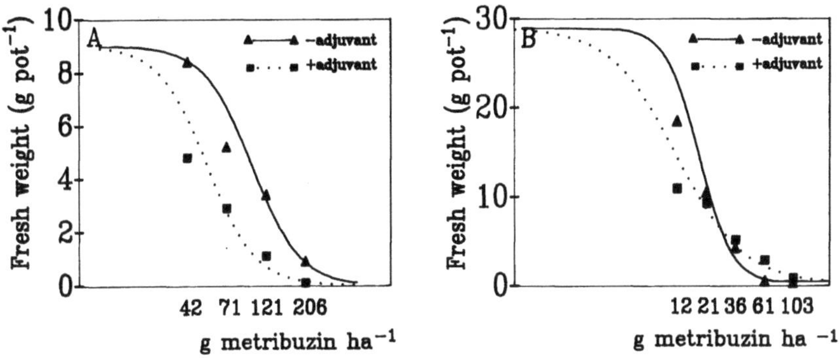

Fig. 4. Relationship between fluazifop-*p*-butyl activity on barley and increasing doses of NPE8 estimated by applying Eq. 5 (**A**) and Eq. 7 (**B**). (Reprinted with permission from Kudsk and Streibig 1993; copyright CRC Press, Boca Raton)

Both functions adequately describe the relationship between herbicide efficacy and adjuvant dose. A significant reduction in activity was observed when increasing the NPE8 concentration from 1.28% to 2.56% (v/v). This was probably due to a phytotoxic effect of the surfactant at the very high dose, inhibiting fluazifop-*p*-butyl uptake and/or translocation in the plants. Thus, the highest adjuvant dose was left out when estimating the curves because both models are monotonic functions and cannot describe a reduction in herbicide activity at high adjuvant doses. Using the *m* parameters estimated from Eq. 5, the JAR was calculated to be 3.17, indicating that at high doses of NPE8 it is possible, without loss of efficacy, to reduce the herbicide dose to approximately one-third of the dose required when fluazifop-*p*-butyl is applied without surfactant.

The *D* parameter denotes the upper limit of the sigmoid curve in Fig. 4B and therefore expresses the maximum activity at high NPE8 doses. Thus, the *D* parameter is comparable to Hewlett's JAR. The *D* parameter was estimated to be 3.39, with a 95% confidence interval of 2.93–3.85, i.e., not significantly different from Hewlett's JAR. In addition to the *D* parameter, the *b* parameter also has a biological significance. The *b* parameter is proportional to the slope around the point of inflection and describes the maximum rate of increase in herbicide activity to small increments in adjuvant dose. The corresponding dose to the point of inflection is a kind of ED_{50} for the adjuvant.

When comparing the two types of curve, it is obvious that the logistic one is much easier to interpret. The maximum potency at high adjuvant doses can be read directly, and the slope of the curve gives a good impression of the interval in which changes in adjuvant dose can be expected to result in significant changes in herbicide activity. In contrast, Hewlett's hyperbola provides little information about the rate of increase in herbicide activity when increasing adjuvant dose, and the maximum obtainable activity must be calculated and cannot be read directly.

In addition to the above example, we have experience only with glyphosate and a tallow amine polyethoxylate (Kudsk and Streibig 1993). The application of the logistic curve therefore needs more validation before it can be generally acceptable. If the logistic curve turns out to be applicable, it provides a better understanding of the relationship between adjuvant dose and herbicide efficacy. Because of the decreasing activity often observed at very high adjuvant doses it may be necessary to leave out these observations, irrespective of whether the

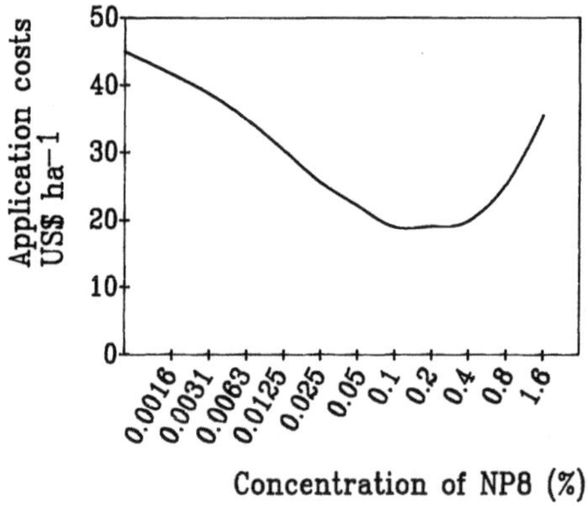

Fig. 5. Application costs (herbicide plus adjuvant) of fluazifop-*p*-butyl with increasing doses of NPE8. The relationship between costs and NPE8 dose was calculated by estimating the relative potencies (Eq. 7), using prices of US$ 45 and 13.5 per liter herbicide and surfactant and assuming that the recommended dose of the herbicide without surfactant was 1.0 l/ha. (The herbicide formulation was an EC formulation containing 125 g fluazifop-*p*-butyl.) (Reprinted with permission from Kudsk and Streibig 1993; copyright CRC Press, Boca Raton)

rectangular hyperbola or the logistic model is used. In any case, this part of the curve is of little practical interest.

Another advantage of first applying the concept of parallel dose-response curves to estimate the relative potencies of the adjuvant doses and then using the relative potencies to estimate the relationship between adjuvant dose and herbicide efficacy is that, as mentioned previously, conclusions are independent of dose and response levels. Furthermore, as differences are expressed quantitatively, results can be used to calculate the most cost-effective combination of herbicide and adjuvant. The application costs for herbicide and surfactant have been calculated using the estimated parameters, current prices of the products on the Danish market, and assuming a recommended dose of 125 g/ha fluazifop-*p*-butyl (1 l/ha) without adjuvant (Fig. 5). The most cost-effective mixture was obtained when the dose of the surfactant was kept between 0.1% and 0.4%. The recommended dose in Denmark is 0.1%. It is likely that on other plant species other surfactant doses may prove to be more cost-effective.

8.4.3 Adjuvants as Modifiers Under Adverse Application Conditions

There are numerous publications on the interaction between application conditions, in particular environmental conditions, and adjuvants on herbicide activity (e.g., Doran and Anderson 1975; Nalewaja et al. 1975; Turner 1981; Langelüddeke et al. 1988). An inherent problem in assessing adjuvant effects under contrasting conditions and using preset dose rates is that the herbicide efficacy can be affected by both adjuvant and application conditions. If, for example, addition of an adjuvant reduces the detrimental effect of rain on herbicide activity, is this a real effect of the adjuvant or simply one due to the increased activity of the herbicide masking the effect of rain, i.e., a dose-response effect? Many of the studies on the interactions between adjuvants and environment have simply not been designed to answer the question which they pose because only one or two herbicide dose rates were included, and the doses with and without adjuvant did not produce equivalent effects. The problem can be overcome either by including more doses and apply nonlinear regression techniques or by selecting doses which produce

Table 2. Relative potencies of tribenuron methyl on *Lamium hybridum* when applied alone and in admixture with 0.05% AE6 at three climates

Climate	Surfactant	Relative potency	95% CI	Relative potency	95% CI[x]
1	None	1.0			
	0.05% AE6	1.47	1.20–1.73	1.47	1.20–1.73
2	None	0.40	0.32–0.49		
	0.05% AE6	0.67	0.52–0.82	1.68	1.30–2.05
3	None	0.48	0.38–0.57		
	0.05% AE6	0.75	0.59–0.91	1.56	1.23–1.90

At climate 1, mean temperature was 5°C, with minimum and maximum temperatures of 2° and 8°C, respectively. At climates 2 and 3, mean temperature was 20°C, with minimum and maximum temperatures of 16° and 24°C at climate 2, and 12° and 28°C at climate 3.
Test for lack of fit: $F_{5,75} = 0.18$
[x]CI, confidence interval

equivalent effects, as was done by Behrens and Elakkad (1981) in their study on rainfastness of two 2,4-D formulations. In this section we present some examples of the way in which the concept of parallel dose-response curves can be applied to assess the influence of adjuvants on the effects of environmental factors on herbicide activity.

One of our research projects is examing the influence of environment on the efficacy of foliage-applied herbicides in climate simulators which can accurately simulate natural diurnal fluctuations in temperature and humidity (Kristensen 1992; Kudsk and Kristensen 1992). Some experiments have included adjuvants to examine the influence of adjuvants on herbicide activity under contrasting climatic conditions. Table 2 presents results from a study on the activity of tribenuron-methyl alone and in admixture with 0.05% AE6 on *L. hybridum* at three climates. Tribenuron-methyl activity was significantly higher at the low temperature regime than at the two high temperature regimes; addition of surfactant improved activity at all three climates. By expressing the relative potency of the surfactant treatments relative to the effect of the corresponding treatments without surfactant it was possible to directly compare the influence of the surfactant under the three climates. Surfactant-induced

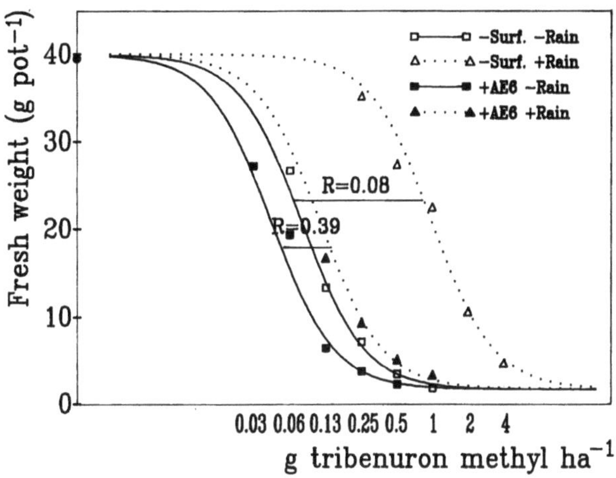

Fig. 6. Dose-response curves and mean fresh weights of *Lamium hybridum* after application of tribenuron-methyl alone and in admixture with 0.05% AE6 without rain or with 3 mm rain 2 h after application. Relative potencies were calculated by dividing the relative potencies of the rain treatments by the corresponding no rain treatments.

enhancement of tribenuron-methyl activity tended to be somewhat lower at the low temperature regime, but differences between climates were not significant, suggesting that the promotion of tribenuron-methyl by 0.05% AE6 is independent of environmental conditions.

Another experiment studies the effect of AE6 on the rainfastness of tribenuron-methyl on *L. hybridum* (Fig. 6). The surfactant reduced the magnitude of the horizontal displacement caused by 3 mm rain 2 h after application. By comparing the rain treatments with the corresponding no rain treatments in a similar way as in Table 2, it was possible to assess the influence of the surfactant on rainfastness. The calculated relative potencies are shown in Fig. 5. Addition of the surfactant did not provide full rainfastness but significantly increased the relative potency from 0.08 to 0.39. A relative potency of 0.08 indicates that if 3 mm rain occurs 2 h after application, the effect of 1 g/ha tribenuron methyl is equivalent to the effect of 0.08 g/ha without rain.

The examples above clearly show the superiority of the concept of parallel dose-response curves when assessing interactions between adjuvants and environment compared to analyses of variance. In contrast to an analysis of variance, this method can provide an answer to the questions put forward in such studies because it is possible to distinguish between the "true" effects of formulations/adjuvants and the effect caused be a shift in response level. The applicability of this concept to assessing interactions between adjuvants and environment or other application factors is well documented (e.g., Kudsk 1989a, 1992; Kudsk and Kristensen 1992; Kudsk and Mathiassen 1994).

8.5 Conclusions

In the future formulation and adjuvants will become even more important. The increasing costs, particularly for development and registration will reduce the number of new active ingredients, and at the same time many of the old compounds are being banned primarily for toxicological and ecotoxicological reasons. The end result will be a reduction in the number of herbicides particularly for use in minor crops. It is therefore necessary to explore the possibilities for improving the activity and selectivity of the available compounds. Formulation and adjuvants are important in achieving this goal.

Another reason why formulation and adjuvants could become even more important is the increasing number of cases of herbicide resistance. A well-designed formulation or an efficient adjuvant could promote the foliar activity of residual herbicides. This in turn could reduce the required dose rate and thereby perhaps delay the evolution of resistant weed populations by reducing the selection pressure.

The political demands of some northern European countries for a substantial reduction in the use of herbicides also puts formulation and adjuvants into a new perspective (Kudsk 1989b). The recommended dose often produces an effect far out on the flat part of the dose-response curve, and the goal of much of the research on reduced doses is to determine the dose that results in an effect near the point at which the dose-response curve flattens out. The risk of reducing the dose is that variability in performance is increased, because at the point where the dose-response flattens out, even small changes in the effective dose, for example, those caused be adverse climatic conditions, will affect herbicide efficacy significantly. Although the average effect of the recommended dose and reduced doses may be similar, variability is often higher with reduced doses. As demonstrated in Sect. 8.4.3, formulation and adjuvants can overcome the impediments imposed by environment and other application factors, and proper selection of formulation and appropiate adjuvants could therefore reduce variability in field performance of reduced herbicide doses.

The selection of formulation and adjuvants is still largely a matter of trial and error, and it is imperative to use a meaningful assay system. We suggest that several doses covering the whole range from no effect to complete kill should be used, and that data be analyzed by nonlinear regressions. As shown in the previous sections, parallel dose-response curves have proven to be a reasonable assumption in most cases. In studies assessing the influence of formulation and adjuvants on herbicide efficacy the concept of parallel dose-response curves should be tested for ease of interpretation. In contrast, in studies examining the effects of formulation or adjuvants on the influence of environment, application techniques, or other variable factors nonlinear regression techniques should be applied, not only for ease of intepretation but also because it is imperative to derive general conclusions. If dose-response curves are not parallel (e.g., Fig. 3C), nonlinear regressions should still be applied. Then relative potencies will no longer be independent of dose and response level, and it will be necessary to estimate relative potencies at more response levels. Nevertheless, nonlinear regression provides results that are much easier to interpret and implement than those from of analysis of variance.

Acknowledgements. The authors gratefully acknowledge the excellent technical assistance of Willy Rasmussen, the financial support of Bayer Denmark A/S, Du Pont de Nemours (Agro) A/S, and Zeneca Agro, and the gift of adjuvants from Joergen Lorentzen A/S.

References

Bayer DE, Drever HR (1965) The effects of surfactants on efficiency of foliar-applied diuron. Weeds 13:222–226

Becher P, Becher D (1969) The effect of hydrophile-lipophile balance on contact angle of solutions of nonionic surface-active agents. In: Van Valkenburg JW (ed) Pesticide formulations research: physical and colloidal aspects. American Chemical Society, Washington DC, pp 15–23 (Advances in chemistry series 86)

Behrens R, Elakkad MA (1981) Influence of rainfall on the phytotoxicity of foliarly applied 2,4-D. Weed Sci 29:349–355

Dexter AG, Burnside OC, Lavy TL (1966) Factors influencing the phytotoxicity of foliar applications of atrazine. Weeds 14:222–228

Doran DL, Anderson RN (1975) Effects of simulated rainfalls on bentazone activity. Weed Sci 23:105–109

Fedtke C (1982) Biochemistry and physiology of herbicide action. Springer, Berlin Heidelberg New York

Finney DJ (1978) Statistical method in biological assay, 3rd edn. Griffin, London

Finney DJ (1979) Bioassay and the practice of statistical inference. Int Stat Rev 47:1–12

Foy CL (1989) Adjuvants: terminology, classification, and mode of action. In: Chow PNP, Grant CA, Hinshalwood AM, Simundsson, E (eds) Adjuvants and agrochemicals, vol II. CRC Press, Boca Raton, pp 1–13

Green JM, Streibig JC (1993) Herbicide mixtures. In: Streibig JC, Kudsk P (eds) Herbicide bioassays. CRC Press, Boca Raton, pp 117–135

Hewlett PS (1979) Measurement of the potencies of drug mixtures. Biometrics 25:477–487

Jansen LL (1964) Relation of structure of ethylene oxide ether-type nonionic surfactants to herbicidal of water soluble herbicides. J Agric Food Chem 12:223–227

Jansen LL (1965) Effects of structural variations in ionic surfactants on phytotoxicity and physical-chemical properties of aqueous sprays of several herbicides. Weeds 13:117–123

Jansen LL (1973) Enhancement of herbicides by silicone surfactants. Weed Sci 21:130–135

Jansen LL, Gentner WA, Shaw WC (1961) Effects of surfactants on the herbicidal activity of several herbicides on aqueous spray systems. Weeds 9:381–405

Keeney FN, Noveroske RL, Flaim TD (1988) Influence of adjuvants on the postemergence phytotoxicity of haloxyfop methyl herbicide. In: Cross B, Scher HB (eds) Pesticide formulations, innovations and developments. American Chemical Society, Washington DC, pp 102–110 (ACS symposium series 371)

Kristensen JL (1992) Research equipment for laboratory experiments with herbicides. Proc 1st Int Weed Contr Congr 2:261

Kudsk P (1989a) Herbicide rainfastmess on indoor and outdoor pot-grown plants. Aspects Appl Biol 21:133

Kudsk P (1989b) Experiences with reduced herbicide doses in Denmark and the development of the concept of factor-adjusted doses. Proceedings of the Brighton Crop Protection Conference-Weeds, pp 545–554

Kudsk P (1992) The effect of adjuvants on the rainfastness of thifensulfuron and tribenuron. In: Foy CL (ed) Adjuvants for agrochemicals. CRC Press, Boca Raton, pp 441–448

Kudsk P, Kristensen JL (1992) Effect of environmental factors on herbicide performance. Proc 1st Int Weed Contr Congr 1:173–186

Kudsk P, Mathiassen SK (1994) Effects of broadleaf herbicides on imazamethabenz-methyl performance on wild oat (Avena fatua L). Weed Res 34 (in press)

Kudsk P, Streibig JC (1993) Formulations and adjuvants. In: Streibig JC, Kudsk P (eds) Herbicide bioassays. CRC, Boca Raton, pp 99–116

Langelüddeke P, Baedelt H, Bieringer H (1988) Trials on the influence of air humidity and rainfall on the efficacy of glufosinate-ammonium. Proceedings of the EWRS symposium, factors affecting herbicide activity and selectivity, pp 227–232

Manthey FA, Nalewaja JD, Szelezniak EF (1989) Esterified seed oils with herbicides. In: Chow PNP, Grant CA, Hinshalwood AM, Simundsson, E (eds) Adjuvants and agrochemicals, vol II. CRC Press, Boca Raton, pp 139–148

Marth PC, Davis FF, Mitchell JW (1945) Herbicidal properties of 2,4-dichlorophenoxyacetic acid applied in dusts containing hygroscopic agents. Bot Gaz 107:131–136

McWhorter CG (1963a) Effects of surfactants on the herbicidal activity of foliar sprays of diuron. Weeds 11:265–269

McWhorter CG (1963b) Effcets of surfactant concentration on Johnsongrass control with dalapon. Weeds 11:83–86

Mitchell JW, Hamner CG (1944) Polyethylene glycols as carriers for growth-regulating substances. Bot Gaz 105:474–483

Nalewaja JD, Pudelko J, Adamczewski KA (1975) Influence of climates and additives on bentazon. Weed Sci 23:504–507

Rudemo M, Ruppert D, Streibig JC (1989) Random-efffects models in nonliniar regression with applications to bioassay. Biometrics 45:349–362

Smith LW, Foy CL, Bayer DE (1966) Structure-activity relationships of alkylphenol ethylene oxide ether non-ionic surfactants and three water-soluble herbicides. Weed Res 6:233–242

Stock D, Holloway PJ, Grayson BT, Whitehouse P (1993) Development of a predictive uptake model to rationalize selection of polyoxyethylene surfactant adjuvants for foliage-applied agrochemicals. Pestic Sci 37:233–245

Streibig JC, Rudemo M, Jensen JE (1993) Dose-response curves and statistical models. In: Streibig JC, Kudsk P (eds) Herbicide bioassays. CRC Press, Boca Raton, pp 29–55

Turner DJ (1981) The effect of additives on control of agropyron repens with glyphosate. Proceedings or the AAB conference on grass weeds in cereals in the United Kingdom, pp 167–175

Turner DJ, Loader MPC (1980) Effect of ammonium sulfate and other additives upon the phytotoxicity of glyphosate to agropyron repens (L) beauv. Weed Res 20:139–146

Wyrill JB III, Burnside OC (1977) Glyphosate toxicity to common milkweed and hemp dogbane as influenced by surfactants. Weed Sci 25:275–287

9 Methodology for Determining Foliar Penetration of Herbicides with Reference to Oil-Based Adjuvants

C. Gauvrit

9.1 Introduction

Oils are used extensively in pesticide formulation and in pesticide application as spray adjuvants. They improve the biological efficacy of several foliage-applied herbicides without impairing their selectivity. Beneficial effects were first observed with atrazine in maize (*Zea mays*; Jones and Anderson 1968; Nalewaja 1968) and with phenmedipham in sugar beet (*Beta vulgaris*: Miller and Nalewaja 1973). More recently, oils have met with success as spray adjuvants for graminicides (Buhler and Burnside 1984; Nalewaja 1986; Barrentine and McWorther 1988) and for bentazone (Doran and Andersen 1975; Nalewaja et al. 1975). Since they allow the use of reduced herbicide application rates, oils also improve ecological compatibility. This is particularly true for vegetable oils, which are more biodegradable than petroleum oils (Cornish et al. 1993).

Among the parameters of herbicidal action, penetration through the plant cuticle seems to be the most affected by oils (Nalewaja and Adamczewski 1977; Nalewaja and Skrzypczak 1986a,b; Schott et al. 1991). However, oils also influence droplet spreading on the plant surface (McWorther and Barrentine 1988) and the physical state (Darchy et al. 1990) and distribution (Hess and Falk 1990) of the active ingredient in the deposit. Some authors also postulate that oils can dissolve epicuticular wax (McWorther and Barrentine 1988). However, the mechanisms and relative importance of these effects are still unclear, and it is obvious that a better understanding of them would help in the design of efficacious adjuvant oils.

There are a number of methodological problems in studying oil-based adjuvants. Some of these arise from the physical properties of oils, some are inherent to studies on cuticular transfer, while others stem from the frequent conflict between two equally worthy aims: (a) to operate under realistic (close to field-use) conditions so as to obtain information of practical value, and (b) to select experimental procedures suitable for mechanistic studies.

9.2 Methodological Problems

9.2.1 Emulsifier

Since oils are practically insoluble in water, they need an emulsifier to be used in aqueous sprays. If the objective is to adhere to field conditions, the effects of oils should be studied in the presence of their emulsifying system. This is usually composed of two or more surfactants, each one being far from pure and often composed of a mixture of isomers and oligomers. Some emulsifiers have phytotoxic effects and, as oils, they can increase droplet spreading and pesticide penetration (Holloway et al. 1992), although it is not a general rule (Schott et al.

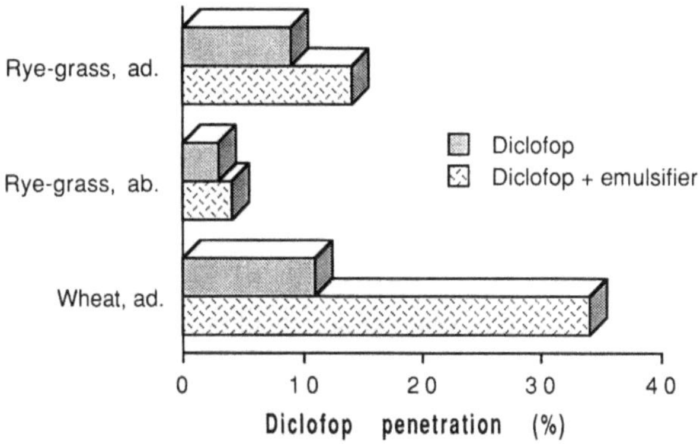

Fig. 1. Influence of emulsifier on the penetration of ^{14}C-diclofop-methyl into wheat and ryegrass. *ab.*, Abaxial surface; *ad.*, adaxial surface. The emulsifier was 25% calcium alkylarylsulphonate, 75% polyethoxylated nonylphenol, 6 mol ethylene oxide, concentration 0.05% (w/v). Penetration was measured 72 h after treatment. (Adapted from Schott et al. 1991)

1991), and depends on plant species (Fig. 1). The situation is even more complex if the herbicide being studied is used as an emulsifiable concentrate, since interactions between the two emulsifying systems cannot be excluded. Hence, although it may be of practical value to study the effects of oils under conditions related to field use, interpretation of results may be obscured by the presence of as many as ten different surfactants. For example, the spreading of 1.25 g/l petroleum and vegetable oils on *Sorghum halepense* leaves was studied using commercial preparations (McWorther and Barrentine 1988). More spreading occurred with a petroleum oil but no information was given on the emulsifiers; these were probably not the same for all the products and may have contributed to the observed differences in spreading.

Complications from emulsifiers can be averted by applying the oil dissolved in a solvent. However, many of the good solvents for oils, such as hexane or toluene, are also efficacious at extracting cuticle constituents. Acetone, however, is useful, at least for oils not consisting of long-chain alkanes (Nalewaja and Skrzypczak 1986a,b; Gillespie et al. 1988; Grafstrom and Nalewaja 1988). Provided that the stomata are closed, acetone appears to exert no toxic action, evaporating in a few seconds from the plant surface with no significant cuticular penetration. In these experiments acetone-water (2:3) was used, and the oil solution was applied prior to herbicide deposition. An even simpler system consists of administering both oil and herbicide as solutions in acetone (Urvoy and Gauvrit 1991; Bravais et al. 1993; Serre et al. 1993).

The part played by the emulsifier in oil action has been checked in penetration studies. The influence of oil on sethoxydim and fluazifop-butyl penetration into oats (*Avena sativa*) was similar whether it was administered emulsified in water or dissolved in acetone (Nalewaja and Skrzypczak 1986a,b). However, although 5% emulsifier was enough for emulsion stability, increasing its concentration to 15% enhanced herbicide penetration. In addition, it suppressed the difference in effectiveness between petroleum and soybean (*Glycine max*) oils (Nalewaja and Skrzypczak 1986b) and between sunflower (*Helianthus annuus*) and linseed (*Linum usitatissimum*) oils (Manthey et al. 1992a). Moreover, in some experiments the action of the emulsifier alone on herbicide penetration was equivalent to that of several oils (Nalewaja and Skrzypczak 1986b; Gillespie et al. 1988).

Hence, the role of the emulsifier must be kept in mind when commercial products are used, and the appropriate checks should be made.

9.2.2 Oil Composition

The oils used as adjuvants are not pure compounds, whether of petroleum or of seed origin. They are usually physically well defined (density, viscosity, flash point, distillation range for petroleum oils), but their chemical composition is rarely disclosed, apart from the content of aromatic compounds or of the unsulfonatable residue. This is a serious inconvenience for mechanistic studies. The components of vegetable oils are more closely related than those of petroleum oils. In the first case, when refined, they differ in fatty acid chain length and degree of unsaturation. On the other hand, beside as low as possible amounts of aromatics, petroleum oils contain a variety of components: linear alkanes, branched alkanes, and the so-called naphthenes, each of these classes having members with differing numbers of carbon atoms. The properties of petroleum oils probably result from the appropriate mixture of these different compounds, but for mechanistic studies, the action of the individual elements have to be ascertained. Attempts have been made to study the action of purified components of vegetable oils and of their alkyl ester derivatives (Urvoy et al. 1992; Bravais et al. 1993; Serre et al. 1993). These are discussed in Sect. 9.3.4.

Oil purity poses another question. Usually, oil concentration is given in grams per liter or refers to liters per hectare (Nalewaja and Skrzypczak 1986a,b). Molarity cannot be calculated since adjuvant oils are not pure compounds. As long as the adjuvants studied have similar molecular weights, this is not a serious problem. However, if one compares seed oils and their methyl ester derivatives, the same concentration in weight reflects a molar ratio of about 3 since, for example, the molecular weights of triolein and methyl oleate are 884 and 298, respectively. This is not so important with regards to concentration per se since studies are usually carried out by depositing on test leaves droplets that rapidly dry down. Consequently, concentration is variable and rapidly increases to have little meaning when the deposit has lost most of its water and other solvents. However, it may be important if one considers the molecular ratio pesticide: oil or the molar surface density of the oil on the cuticle.

9.2.3 In Situ or Isolated Cuticles?

The various methods of studying foliar penetration of pesticides have recently been discussed (Schreiber and Schönherr 1992), and we describe here the points which are most relevant to the study of oil action. Isolated cuticles are a valuable tool to understand the mechanisms involved in foliar penetration. They are simpler systems than in situ cuticles since they are not exposed to the physical and physiological influences of the living tissue. Sorption, diffusion, and desorption of xenobiotics can be studied under well-defined conditions, with a greater freedom to impose variables. For these reasons, quantitative data, for example, partition, diffusion and permeability coefficients can be obtained, as well as information on the influence of factors such as lipophilicity and molar volume (Schönherr and Riederer 1989).

However, as far as adjuvant oils are concerned, there are some limitations to the use of isolated cuticles. Firstly, only thick (and astomatous) cuticles, which can sustain the mechanical strains of isolation,

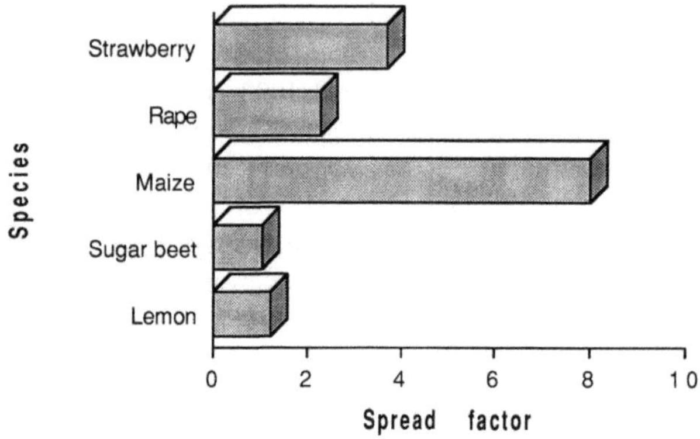

Fig. 2. Spread factors for droplets of Risella oil (with 10% dichloromethane) sprayed onto leaves of various species. Spread factor = $R_d^2 : R_f^2$, where R_f is the radius of the in-flight droplet and R_d is the radius of the dried deposit. (Adapted from Baker et al. 1983)

can be obtained without imperfections. Unfortunately, most of the species (crops or weeds) of economic importance have thin and delicate cuticles. For example, the cuticle of sugar beet and maize are 0.2 and 0.01 um thick, respectively (Stevens and Baker 1987). Hence, studies with isolated cuticles rely on material which can be very different in structure and composition from cuticles of practical interest. This is a drawback since the influence of adjuvant oils seems in some cases to be species-dependent (Manthey et al. 1992b). The cuticles of Gramineae (crops and weeds) are generally thin; this is unfortunate since oils are particularly effective on graminicide activity.

Secondly, during the preparation, isolated cuticles are shaken, even tumbled, for hours in liquid solutions. If crystalline epicuticular wax is present, it is very probably damaged. This is of utmost importance in studies on oil since, as for other lipophilic compounds (Holloway and Silcox 1985), the behavior and effect of oils is influenced by the presence of crystalline epicuticular wax. The spreading of oil droplets was restricted on the smooth surfaces of sugar beet and lemon (*Citrus limon leaves*), whereas they readily spread on those of rape (*Brassica napus*), maize, and strawberry (*Fragaria ananassa*), whose epicuticular wax is crystalline (Fig. 2, Baker et al. 1983). The increase in the foliar penetration of diclofop-methyl brought about by methyl oleate was higher in glaucous maize than on glossy maize, which possess crystalline and amorphous epicuticular wax, respectively (Urvoy et al. 1992). Hence, as far as oils are concerned, one cannot be satisfied with procedures utilizing only isolated cuticles. Intact plants should also be used; several species of differing cuticular types need to be selected.

Studies on intact plants generally involve depositing droplets containing the labeled pesticide with or without the oil under examination, as described in Sect. 9.3.4. Another recently described method uses detached leaves submersed in aqueous solutions of ^{14}C-labelled pesticides (Schreiber and Schönherr 1992). The kinetics of uptake could be precisely established, as well as the compartmentation of the absorbed pesticide. Of course, under such conditions pesticide absorption takes place from a liquid phase, whereas after a spray most of it occurs from a dried deposit (Stevens et al. 1988). Hence, this method cannot take into account phenomena linked with the deposit such as spreading, distribution and physical state of the active ingredient, and diffusion from a (semi)solid phase.

9.3 Methodologies

9.3.1 Spreading

The simplest method of studying droplet spreading on plant surfaces uses the light microscope (Baker et al. 1983; McWorther and Barrentine 1988; Schott et al. 1991). Incorporation of fluorescent tracers (hydrophilic or lipophilic) in the spray liquid give clear pictures of the deposits upon illumination with UV light. Scanning electron microscopy can also be used (Boize et al. 1976; Baker et al. 1983), sometimes with X-ray analysis. Autoradiography offers the same advantage, but some caution must be exerted since it detects not only the radioactive pesticide or oil in the deposit but also the label that has penetrated the plant. If it is mobile, the image of the deposit after several hours may be blurred. An elegant solution has been to make autoradiographs of cellulose acetate films stripped from the plant surface (Baker et al. 1983).

The use of such techniques has led to interesting results as regards the dependence of spreading on oil composition and on cuticular type.

Firstly, paraffinic oils spread to completely cover *S. halepense* leaves, whereas soybean oil covered only 30%–50%. Droplets containing 1.25% paraffinic or soybean oil emulsified in water spread on *S. halepense* leaves, the effect being greater with the former (McWorther and Barrentine 1988). Differences in spreading behavior were observed between purified triolein and methyl oleate when they were also deposited as solutions (without emulsifier) in acetone-water (19:1) on pea leaves (Bravais et al. 1993). Droplets containing methyl oleate spread to a greater extent, as did droplets containing a petroleum oil of unknown composition.

Secondly, extensive spreading of oil droplets was observed only on leaves with crystalline epicuticular wax deposits such as maize, strawberry and rape, as opposed to sugar beet and lemon leaves (Baker et al. 1983; Fig. 2). Droplets containing formulated diclofop spread on ryegrass (*Lolium multiflorum*) leaves when oil (petroleum, rapeseed, or methylated rapeseed) was present, but only on the adaxial surface which is covered with crystalline epicuticular wax. On the abaxial surface (covered with amorphous epicuticular wax) no spreading was observed (Schott et al. 1991).

In studies on oil adjuvants, spreading must be examined because it reflects the affinity of the oil for the plant surface. However, its importance in oil action should not be overestimated since the influence of deposit area on pesticide penetration is not verified (Price 1982; Holloway and Silcox 1985; Wanamarta et al. 1989).

9.3.2 Effect on Wax

After treatment with undiluted paraffinic oil the structure of the epicuticular wax on *S. halepense* leaves was smoothed and cracks appeared on the amorphous layer at the base (McWorther and Barrentine 1988). Although the authors claimed that the wax was dissolved, it may have been only melted because oils are generally poor solvents of wax, although they can rapidly impregnate wax deposits (Gauvrit and Cabanne 1993). This was studied on wax removed from green foxtail (*Setaria viridis*) leaves and deposited as a solution in chloroform onto glass slides (Manthey and Nalewaja 1992). After the chloroform had evaporated, an oil droplet was placed on the wax. The slide was tilted after

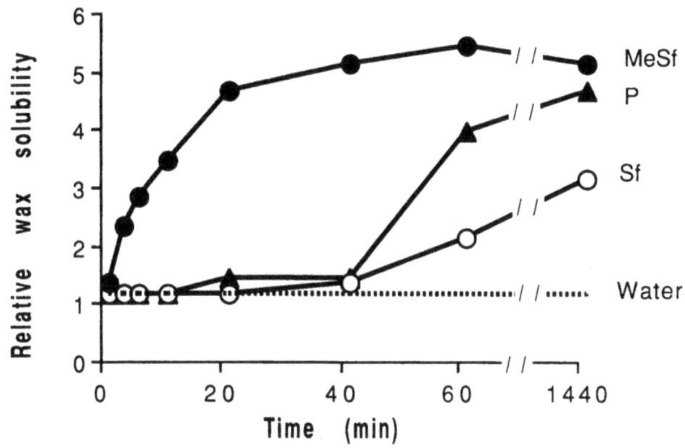

Fig. 3. Relative wax solubility ratings for oils in *Setaria viridis* wax. *Sf*, Sunflower oil; *MeSf*, methylated sunflower oil; *P*, petroleum oil. (Adapted from Manthey and Nalewaja 1992)

various times and the wax was then firmly wiped perpendicular to the oil drainage. A relative wax solubility number was assigned on a rating scale ranging from 1 (no solubility) to 9, according to the shape and the appearance of the spot were the oil droplet had been placed. A comparison between petroleum oil, sunflower oil, and its methyl ester is shown in Fig. 3. Relative wax solubility of the latter increased rapidly, whereas with petroleum and sunflower oil a 40 min time-lag was observed. After 24 h petroleum oil had a relative wax solubility number close to that of the ester derivative but that of sunflower oil was significantly lower. Similar values were obtained with cotton (*Gossypium hirsutum*) and linseed oils. It is interesting to note that relative wax solubility ratings ranked the differing oils in the same order as their enhancement of herbicide penetration (Nalewaja and Skrzypczak 1986a,b).

Phenomena such as the melting of wax by oils and solubilization of oils in wax are probably crucial in explaining the increase in the diffusion of pesticides through the cuticle. The interaction of oils with cuticular wax deserves more quantitative studies.

9.3.3 Physical State and Distribution of Active Ingredient

Oils probably affect the physical state of pesticides which are soluble in them, for instance 2,4-D esters (Worthing and Hance 1991) and fluazifop-butyl (Nalewaja and Skrzypczak 1986a,b). On the other hand, rapeseed oil inhibited the recrystallisation of phenmedipham (Darchy et al. 1990). Both actions may result in higher penetration rates (Price 1982).

The distribution of herbicide in a deposit can be altered by oils, as shown on soybean leaves with scanning electron microscopy and cathodoluminescence. For example, when [(4-chloro-*o*-tolyl)oxy]acetic acid (MCPA) was applied in a water-based spray, most of the herbicide was present in the depressions above the anticlinal walls, but when petroleum oil was added to the spray solution, MCPA was uniformly distributed over both the anticlinal and periclinal cell walls (Hess and Falk 1990).

9.3.4 Penetration

9.3.4.1 General Procedures

The effect of oils on the penetration of herbicides has been studied using conventional methods. The ^{14}C-labeled herbicide is applied as droplets on plant leaves. After various times penetration is evaluated by washing the deposit with an appropriate solvent and then by determining the radioactivity present in the wash and in the leaf by scintillation counting. Droplet volumes range from 0.2 μl (Serre et al. 1993) to 15 μl (for example, Nalewaja and Skrzypczak 1986a), which represents 10–20 and 15 $\mu l/cm^{-2}$, respectively, and deposit areas of 1–2 mm^2 and 1 cm^2, respectively.

In most cases, the plant species were chosen according to herbicide use in the field, namely, maize, *Setaria* spp. and *Echinochloa crusgalli* for atrazine (Coats and Foy 1974b; Nalewaja and Adamczewski 1977), *Amaranthus retroflexus*, *Brassica kaber* and soybean for bentazone (Nalewaja and Adamczewski 1977), *Avena sativa* for fluazifop-butyl and sethoxydim (Nalewaja and Skrzypczak 1986a,b; Grafstrom and Nalewaja 1988), ryegrass for diclofop-methyl (Schott et al. 1991), and *Setaria faberi* and *Sorghum bicolor* for primisulfuron (Hart et al. 1992). However, in other studies the properties of epicuticular waxes and of foliar surfaces were the rationale behind the choice of species, rather than their economic importance for the herbicides. For example, pea (*Pisum sativum*, crystalline epicuticular wax) and cleavers (*Galium aparine*, amorphous epicuticular wax) were used with quizalofop-ethyl, in spite of its complete lack of action against dicotyledonous plants (Bravais et al. 1993; Serre et al. 1993).

Nonpenetrated lipophilic herbicides are collected either by washing the deposit area with 0.5 ml acetone then with the same amount of chloroform (for example, Urvoy et al. 1992) or by dipping the leaf ten times in a toluene-based scintillation fluid (e.g., Nalewaja and Skrzypczak 1986a). The latter procedure saves time and solvent and solubilizes both the deposit and the epicuticular wax. Extraction of herbicide located inside the tissue seems to be negligible since penetration percentages as high as 97% of the applied label were recorded (Nalewaja and Adamczewski 1977). When the herbicide was less lipophilic, the leaf was dipped in methanol-water (1:1 with primisulfuron; Hart et al. 1992) or dipped in water before the scintillation fluid (with bentazone; Nalewaja and Adamczewski 1977). When the penetration of purified triolein and esters of oleic acid was studied, the deposit was washed with acetone then chloroform (for example, Bravais et al. 1993).

9.3.4.2 Preparation and Application of Oil

As already mentioned (Sect. 9.2.1), the effects of the surfactants associated with the oils and the herbicides must be taken into consideration. This can be done in various ways. (a) Herbicide in solution and formulated oils (Coats and Foy 1974b): atrazine was applied first, dissolved in methanol-water (1:3), then the emulsified oils. (b) Formulated herbicides and oils in solution (Nalewaja and Skrzypczak 1986a,b; Gillespie et al. 1988; Grafstrom and Nalewaja 1988): 5-μl applications of the following were made: oil dissolved in acetone-water (2:3), then formulated herbicide, then radiolabeled herbicide. (c) Both herbicides and oils in solution (Bravais et al. 1993; Serre et al. 1993): 0.5 or 0.2 μl acetone containing the herbicide and the oil under study were applied. In the latter study the acetone contained 5% water to avoid extensive spreading on plant surfaces with crystalline epicuticular wax, such as pea. Under these conditions, droplet spreading was similar on pea and on *G. aparine*, which is covered with amorphous wax. A higher water content

Table 1. Concentrations of herbicide and oil adjuvants used in foliar penetration studies and the resulting molar ratios

Herbicide	Herbicide concentration (m*M*)	Triglyceride concentration (m*M*)	Ester concentration (m*M*)	Molecular ratio	Reference
Bentazone	4.7	8.7[a]	–	1.9	Nalewaja and Adamczewski (1977)
Fluazifop	1.1	4.3[a]	–	3.9	Nalewaja and Skrzypczak (1986a)
Fluazifop	1.0	4.1	–	4.1	Nalewaja et al. (1986)
Fluazifop	1.0	–	11.0	11.0	Nalewaja et al. (1986)
Sethoxydim	1.8	4.3[a]	–	2.4	Nalewaja and Skrzypczak (1986b)
Sethoxydim	1.2	4.1	–	3.4	Nalewaja et al. (1986)
Sethoxydim	1.2	–	11.0	9.2	Nalewaja et al. (1986)
Quizalofop	1.1	4.1	–	3.4	Manthey et al. (1992a,b)
Quizalofop	1.1	–	10.0	10.0	Manthey et al. (1992a,b)
Quizalofop	10.0	10.0	–	1.0	Serre et al. (1993); Bravais et al. (1993)
Quizalofop	10.0	–	10.0	1.0	Serre et al. (1993); Bravais et al. (1993)
Diclofop	10.0	10.0	–	1.0	Urvoy et al. (1992)
Diclofop	10.0	–	10.0	1.0	Urvoy et al. (1992)
Diclofop	2.9	3.4	–	1.2	Schott et al. (1993)
Diclofop	2.9	–	8.9	3.1	Schott et al. (1993)

Table 1. Continued

Herbicide	Herbicide concentration (m*M*)	Triglyceride concentration (m*M*)	Ester concentration (m*M*)	Molecular ratio	Reference
Phenmedipham	10.0	10.0	–	1.0	Serre et al. (1993); Bravais et al. (1993)
Phenmedipham	10.0	–	10.0	1.0	Serre et al. (1993); Bravais et al. (1993)
Primisulfuron	0.19	–	20.2[a]	106.3	Hart et al. (1992)

[a] Technical compound, approximate value. For triglyceride and methyl ester derivatives densities of 0.9 and 0.8, respectively, were taken.

is needed to limit spreading on the leaves of Gramineae but some oils would no longer be soluble.

The concentration of herbicide and oil in the deposit are not defined variables since the water or the solvent evaporate quickly. However, they do indicate the ratio herbicide:oil, which is a critical factor for mechanistic studies. Workers do not usually indicate initial concentrations, but these can be calculated from the given field-related variables such as rate and volume of application. In experiments with bentazone (Nalewaja and Adamczewski 1977) the initial bentazone concentration in the deposit was 4.7 m*M* (0.56, 495 l/ha), whereas the oil amount corresponded to 4.7 l/ha. The composition of the petroleum oil was not indicated, the vegetable oils came from linseed. If they were composed of only one triglyceride with a molecular weight of 880 and had a density of 0.9, the concentration would be 8.7 m*M*. This assumption is reasonable because (a) about 90% of the fatty acids from linseed oil have 18 carbon atoms (Röbbelen 1991), and (b) the molecular weights of triglycerides differing only in their degree of unsaturation are very close since only a few hydrogen atoms are involved. In other words, the molecular ratio between bentazone and the triglycerides was about 1:1.9. Table 1 gives some values from other experiments.

Three points can be made. Firstly, herbicide concentrations between 1 and 10 m*M* are realistic for compounds applied at 100–1000 g/ha since in practice application volume ranges from 70 to 500 l/ha. However, with herbicides active at low rates, such as sulfonylureas, realistic concentrations are well below 1 m*M*. Secondly, applying the same amount (by weight) per hectare of nontransformed vegetable oil and of transesterified vegetable oil results in the application of about three times more of the latter in terms of moles. Finally, oil:herbicide molar ratios vary from 1 (diclofop-methyl, phenmedipham) to more than 100 (primisulfuron), that for quizalofop-ethylis, from 1 to 10. To date, we do not know what influence this ratio has on oil effects. If oils act through some interaction with herbicide molecules, the ratio would certainly have an influence. If oils act indirectly via alteration of the physical properties of the cuticle, the influence may be less. However, even in this case, the comment concerning the relative amounts of nontransformed and transesterified vegetable oils still holds.

9.3.4.3 Comparisons Among Various Oils

The effect of petroleum and nontransformed linseed oils on bentazone foliar penetration into *A. retroflexus* and *B. kaber* was comparable, one of the preparations of linseed oil being even more active than the petroleum oils (Nalewaja and Adamczewski 1977). The herbicide and all oils were formulated, and no information was given on the emulsifiers. With formulated sethoxydim the influence of oils was studied without the emulsifier. Herbicide penetration into oats was generally greater when applied with petroleum than with seed oils, but the difference was small and disappeared when emulsifier was added (Nalewaja and Skrzypczak 1986b). With graminicides of the "fop" series, the increase in penetration brought about by petroleum oils was greater than that of vegetable oils. With formulated fluazifop-butyl (Nalewaja and Skrzypczak 1986a), petroleum oil increased penetration into oats, whereas safflower (*Carthamus tinctorius*), linseed, soybean, and sunflower oils had no action (Fig. 4). The presence of emulsifiers in the oils did not affect their performance. Rapeseed oil was less effective than petroleum oil in enhancing the penetration of formulated diclofop-methyl into ryegrass (Schott et al. 1991) but methylated rapeseed oil was as effective (oils were formulated). Methylated seed oil from an undisclosed source and composition increased the penetration of formulated primisulfuron into *S. faberi* and *S. bicolor* to a greater extent than petroleum oil (Hart et al. 1992).

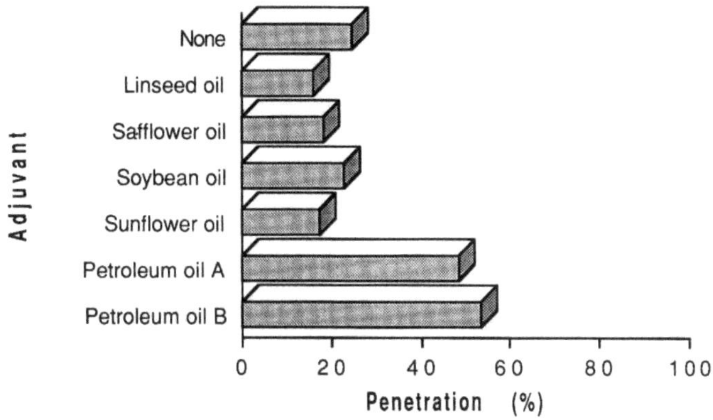

Fig. 4. Penetration of ^{14}C-fluazifop-butyl into oats in the presence of various adjuvants 48 h after treatment. (Adapted from Nalewaja and Skrzypczak 1986a)

Other comparisons between nontransformed and methylated seed oils have been made using purified compounds and reached the same conclusions as far as unsaturated fatty acids were concerned. The herbicides studied were fluazifop-butyl and sethoxydim on oats (Nalewaja et al. 1986), diclofop-methyl on maize (Urvoy and Gauvrit 1991), and quizalofop-ethyl on oats and *S. glauca* (Manthey et al. 1992a). However, the penetration of quizalofop-ethyl into pea and *G. aparine* was not influenced by methyl oleate and triolein (Bravais et al. 1993). This discrepancy may have two explanations: (a) that the herbicide was solubilized in acetone in the latter experiment and emulsified in water in the former, and (b) that the species were different, Gramineae in the first case and dicotyledonous in the second case (even though pea has crystalline epicuticular wax). Although it is not yet possible to decide between these two possibilities, these observations are interesting because the former possibility relates to the formulation problem, and the latter may indicate that there is some specificity of action linked to target species. Indeed, triolein stimulated diclofop-methyl penetration into glaucous maize but had no action on its penetration into glossy maize, whereas methyl oleate stimulated diclofop-methyl penetration into both maize types (Fig. 5, Urvoy et al. 1992).

From such experiments the following picture emerges. Vegetable oils are generally less efficacious than petroleum oils at enhancing

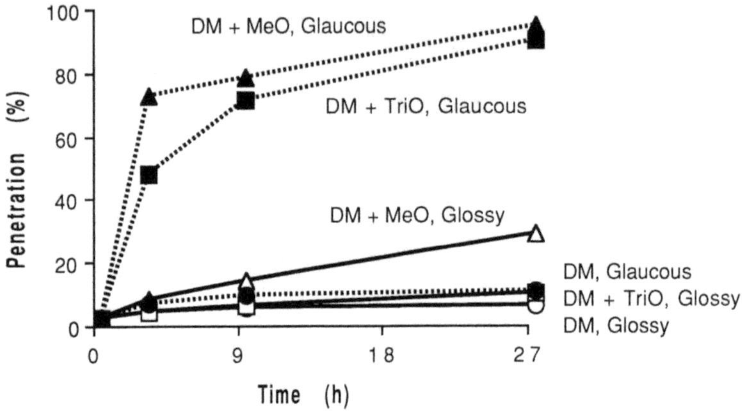

Fig. 5. Penetration of formulated ^{14}C-diclofop-methyl into maize leaf. *DM*, Diclofop-methyl alone; *DM + MeO*, diclofop-methyl + emulsified methyl oleate; *DM + TriO*, diclofop-methyl + emulsified triolein; *Glaucous*, maize with crystalline epicuticular wax; *Glossy*, maize with amorphous epicuticular wax. The emulsifier was the same as in Fig. 1. (Adapted from Urvoy et al. 1992)

herbicide penetration, although their methylated derivatives perform as well. On the whole, no important differences have been observed between seed oils of various origins.

9.3.5 Modification of Oils

Modification of oils serves two purposes, not only to design better adjuvants but also to shed some light on their mode of action. For instance, no difference in atrazine penetration into maize was detected whether it was applied with petroleum oils of the naphthenic type or of the paraffinic type (Coats and Foy 1974b). However, all oils contained more than 20% of aromatic compounds which may have obscured differences between the oil types (Coats and Foy 1974a). Few studies with petroleum oils of well-defined composition have been presented, and they are always rather complex mixtures.

Chemical modification is in principle easy with vegetable oils, due to the reactivity of the fatty acids, of which a great variety are available,

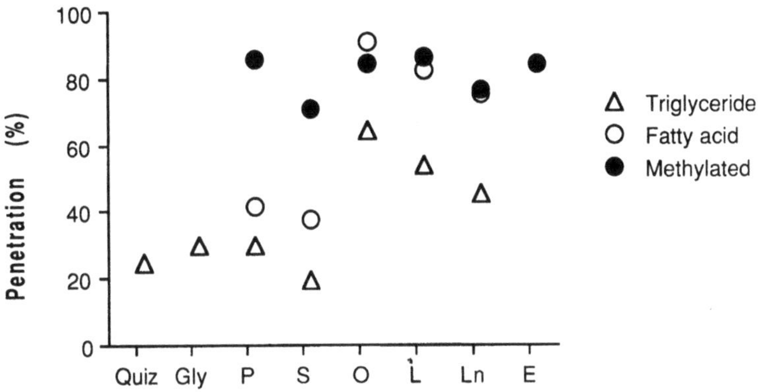

Fig. 6. Penetration of ^{14}C-quizalofop-ethyl into oats 48 h after treatment. *Quiz*, Quizalofop-ethyl alone. The additions were: *Gly*, glycerol; *P*, palmitic acid; *S*, stearic acid; *O*, oleic acid; *L*, linoleic acid; *Ln*, linolenic acid; *E*, erucic acid. (Adapted from Manthey et al. 1992a)

many of them purified. It is possible to compare the effectiveness of purified triglycerides, fatty acids, and fatty acid methyl esters with differing numbers of double bonds on the foliar penetration of herbicides (Nalewaja et al. 1986; Manthey et al. 1992a). The action of fatty acids was similar to that of their methyl esters. As mentioned above, stearic acid derivatives were generally less effective than derivatives of unsaturated fatty acids. With fluazifop-butyl and sethoxydim, no differences among oleic, linoleic, and linolenic-based compounds were found, whereas with quizalofop-ethyl there was a stronger dependence on the number of double bonds, linolenic derivatives often being less efficacious (Fig. 6, Manthey et al. 1992a).

In addition, it is possible to synthesize inexpensive analogues from radiolabeled fatty acids, which permits the study of their behavior on plant surfaces. For instance, methyl oleate penetrated into maize leaves, whatever the structure of its epicuticular wax, whereas triolein did not penetrate into maize leaves with amorphous wax (Urvoy et al. 1992). In addition, in contrast to methyl oleate, it did not promote diclofop-methyl penetration into leaves of this maize type (Fig. 5).

Alkyl esters of oleic acid (methyl, butyl, octyl, dodecyl, and dodecanyl) have been synthesized and their effect on the penetration of phen-

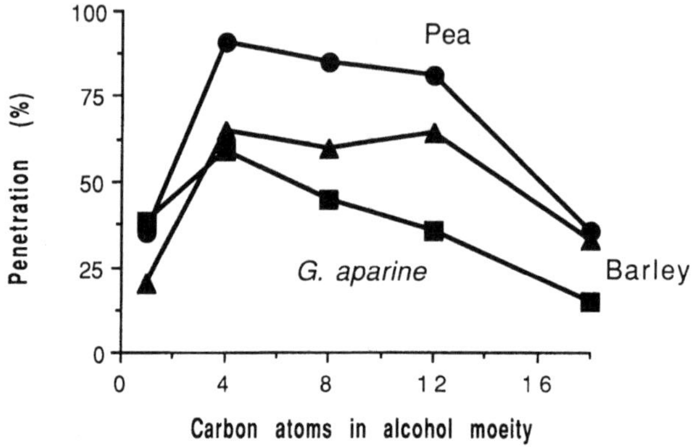

Fig. 7. Penetration of phenmedipham into the leaves of three species in the presence of various esters of oleic acid. All compounds were applied at 10 m*M* in 95% acetone. Measurements were made 27 h after treatment. (Adapted from Serre et al. 1993)

medipham and quizalofop-ethyl studied (Serre et al. 1993). All of them increased cuticle permeability to phenmedipham in pea, *G. aparine* and barley (*Hordeum vulgare*). In all three species the greatest increase was observed with the butyl derivative although the octyl and dodecyl derivatives performed almost equally well (Fig. 7). With quizalofop-ethyl, the influence of the alkyl chain length depended on the plant species. In pea and *G. aparine*, the penetration increase was limited and did not depend on the number of carbon atoms in the alcoholic moeity. In contrast, in barley the relationship was similar to that observed with phenmedipham. It is too early to draw conclusions and give interpretations from these observations, but it is interesting to note that they address several questions of importance in the design of oil adjuvants. What is the optimum carbon chain length for an oil adjuvant? Is the action of oils dependent on plant surface properties? Is the action of oils dependent on the herbicide?

9.4 Conclusions

The mechanisms of oil action are poorly understood, although their effect on herbicide penetration is now well documented. It does not appear to be restricted to lipophilic compounds since oils increase the foliar penetration of compounds such as bentazone and primisulfuron. This may indicate that they do not interact with the herbicide. However, some interaction may exist, at least in a few cases since, for instance, the action of seed oils is different on fluazifop-butyl than on sethoxydim. This example illustrates the need for a better understanding of the physical interactions between herbicides and oils (in particular in the deposit) and between oils and the various types of plant cuticle. Further work is also needed to elucidate structure-activity relationships between oils and pesticides.

Acknowledgements. I am indebted to F. Cabanne for helpful discussions during the preparation of this manuscript.

References

Baker EA, Hunt GM, Stevens PJG (1983) Studies on plant cuticle and spray droplet interactions: a fresh approach. Pestic Sci 14:645–658

Barrentine WL, McWhorter CG (1988) Johnsongrass (sorghum halepense) control with herbicides in oil diluents. Weed Sci 36:102–110

Boize L, Gudin C, Purdue G (1976) The influence of leaf surface roughness on the spreading of oil spray drops. Ann Appl Biol 84:205–211

Bravais F, Cabanne F, Gauvrit C (1993) Influence of trioleine and methyl-, ethyl-, and propyl-oleate on the deposit shape and the foliar penetration of phenmedipham and quizalofop-ethyl. Mededelingen van de Faculteit Landbouwwetenschappen, Rijksuniversiteit, Gent 58/3a:803–807

Buhler DD, Burnside OC (1984) Effect of application factors on postemergence phytotoxicity of fluazifop-butyl, haloxyfop-methyl, and sethoxydim. Weed Sci 32:574–583

Coats GE, Foy CL (1974a) Effects of atrazine-phytobland oil combinations on $^{14}CO_2$-fixation and transpiration. Weed Sci 22:215–220

Coats GE, Foy CL (1974b) Effects of petroleum oils on the uptake of atrazine-^{14}C by corn. Weed Sci 22:220–226

Cornish A, Battersby NS, Watkinson RJ (1993) Environmental fate of mineral, vegetable and transesterified vegetable oils. Pestic Sci 37:173–178

Darchy F, Millet JC, Laverriere PH (1990) Intérêt d'une nouvelle formulation à base de phenmédiphame pour le désherbage de postlevée des betteraves sucrières. Proceedings of the 14th conference of COLUMA, Versailles, pp 693–700

Doran DL, Andersen RN (1975) Effects of simulated rainfall on bentazone activity. Weed Sci 23:105–109

Gauvrit C, Cabanne F (1993) Oils for weed control: uses and mode of action. Pestic Sci 37:147–153

Gillespie GR, Skrzypczak GA, Nalewaja JD (1988) Absorption and translocation of CGA-82725 with additives. Weed Sci 36:282–285

Grafstrom LD, Nalewaja JD (1988) Uptake and translocation of fluazifop in green foxtail (setaria viridis). Weed Sci 36:153–158

Hart SE, Kells JJ, Penner D (1992) Influence of adjuvants on the efficacy, absorption, and spray retention of primisulfuron. Weed Technol 6:592–598

Hess FD, Falk RH (1990) Herbicide deposition on leaf surfaces. Weed Sci 38:280–288

Holloway PJ, Silcox D (1985) Behaviour of three nonionic surfactants following foliar application. Proceedings of the British Crop Protection Conference, Weeds, pp 297–302

Holloway PJ, Wong WW-C, Partridge H, Seaman D, Perry RB (1992) Effects of some nonionic polyoxyethylene surfactants on uptake of ethirimol and diclobutrazol from suspension formulations applied to wheat leaves. Pestic Sci 34:109–118

Jones GE, Anderson GW (1968) Atrazine as a foliar application in an oil water emulsion. Proc North Centre Weed Contr Conf 20:27–29

Manthey FA, Nalewaja JD (1992) Relative wax solubility and phytotoxicity of oil to green foxtail [setaria viridis (L.) Beauv.]. In: Foy CL (ed) Adjuvants for agrochemicals. CRC Press, Boca Raton, pp 463–471

Manthey FA, Szelezniak EF, Anyszka ZM, Nalewaja JD (1992a) Foliar absorption and phytotoxicity of quizalofop with lipid compounds. Weed Sci 40:558–562

Manthey FA, Matysiak R, Nalewaja JD (1992b) Petroleum oil and emulsifier affect the phytotoxicity of imazethapyr. Weed Technol 6:81–84

McWhorter CG, Barrentine WL (1988) Spread of paraffinic oil on leaf surfaces of johnsongrass (sorghum halepense). Weed Sci 36:111–117

Miller SD, Nalewaja JD (1973) Effects of additives upon phenmedipham for weed control in sugarbeet. Weed Sci 21:67–70

Nalewaja JD (1968) Postemergence atrazine treatments for weed control in corn. Proc North Centre Weed Contr Conf 23:12–13

Nalewaja JD (1986) Seed oils with herbicides. Mededelingen van de Faculteit Landbouwwetenschappen, Rijksuniversiteit, Gent 51(2a):301–310

Nalewaja JD, Adamczewski KA (1977) Uptake and translocation of bentazone with additives. Weed Sci 25:309–315

Nalewaja JD, Skrzypczak GA (1986a) Absorption and translocation of sethoxydim with additives. Weed Sci 34:657–663

Nalewaja JD, Skrzypczak GA (1986b) Absorption and translocation of fluazifop with additives. Weed Sci 34:572–576

Nalewaja JD, Pudelko J, Adamczewski KA (1975) Influence of climate and additives on bentazone. Weed Sci 23:504–507

Nalewaja JD, Skrzypczak GA, Gillespie GR (1986) Absorption and translocation of herbicides with lipid compounds. Weed Sci 34:564–568

Price CE (1982) A review of the factors influencing the penetration of pesticides through plant leaves. In: Cutler DF, Alvin KL, Price CE (eds) The plant cuticle. Academic, London, pp 237–252

Röbbelen G (1991) The genetic improvement of seed oil. Chem Indust 7 October, pp 713–716

Schönherr J, Riederer M (1989) Foliar penetration and accumulation of organic chemicals in plant cuticles. Rev Environ Contam Toxicol 108:1–70

Schott J-J, Dufour J-L, Gauvrit C (1991) Effects of adjuvants on herbicidal action. III. Effects of petroleum and rapeseed oils on diclofop-methyl action on ryegrass. Agronomie 11:27–34

Schreiber L, Schönherr J (1992) Analysis of foliar uptake of pesticides in barley leaves: role of epicuticular waxes and compartmentation. Pestic Sci 36:213–221

Serre I, Cabanne F, Gauvrit C (1993) Seed oil derivatives as adjuvants: influence of methyl to octadecyl oleates on the penetration of herbicides through various plant cuticles. Mededelingen van de Faculteit Landbouwwetenschappen, Rijksuniversiteit, Gent 58/3a:795–802

Stevens PJG, Baker EA (1987) Factors affecting the foliar absorption and redistribution of pesticides. I. Properties of their leaf surfaces and their interactions with spray droplets. Pestic Sci 19:265–281

Stevens PJG, Baker EA, Anderson NH (1988) Factors affecting the foliar absorption and redistribution of pesticides. II. Physicochemical properties of the active ingredient and the role of surfactant. Pestic Sci 24:31–53

Urvoy C, Gauvrit C (1991) Seed oils as adjuvants: penetration of glycerol trioleate, methanol oleate and diclofop-methyl in maize leaves. Proceedings of the Brighton Crop Protection ConferenceWeeds, pp 337–342

Urvoy C, Pollacsek M, Gauvrit C (1992) Seed oils as additives: penetration of triolein, methyl oleate and diclofop-methyl in maize leaves. Weed Res 32:375–383

Wanamarta G, Penner D, Kells JJ (1989) Identification of efficacious adjuvants for sethoxydim and bentazone. Weed Technol 3:60–66

Worthing CR , Hance RJ (1991) 2,4-D. In: Worthing CR, Hance RJ (eds) The pesticide manual, 9th edn. British Crop Protection Council, Farnham, p 218

Subject Index

FSC
www.fsc.org
MIX
Papier aus verantwortungsvollen Quellen
Paper from responsible sources
FSC® C105338

If you have any concerns about our products,
you can contact us on
ProductSafety@springernature.com

In case Publisher is established outside the EU,
the EU authorized representative is:
Springer Nature Customer Service Center GmbH
Europaplatz 3, 69115 Heidelberg, Germany

Printed by Libri Plureos GmbH
in Hamburg, Germany